Everyday Mathematics®

The University of Chicago School Mathematics Project

Student Math Journal
Volume 2

Grade 3

Mc Graw Hill **Wright Group**

The McGraw-Hill Companies

The University of Chicago School Mathematics Project (UCSMP)

Max Bell, Director, UCSMP Elementary Materials Component; Director, *Everyday Mathematics* First Edition
James McBride, Director, *Everyday Mathematics* Second Edition
Andy Isaacs, Director, *Everyday Mathematics* Third Edition
Amy Dillard, Associate Director, *Everyday Mathematics* Third Edition

Authors

Max Bell	Amy Dillard	Kathleen Pitvorec
Jean Bell	Robert Hartfield	Peter Saecker
John Bretzlauf	Andy Isaacs	
Mary Ellen Dairyko*	James McBride	

*Third Edition only

Technical Art
Diana Barrie

Teachers in Residence
Lisa Bernstein, Carole Skalinder

Editorial Assistant
Jamie Montague Callister

Contributors
Carol Arkin, Robert Balfanz, Sharlean Brooks, James Flanders, David Garcia, Rita Gronbach, Deborah Arron Leslie, Curtis Lieneck, Diana Marino, Mary Moley, William D. Pattison, William Salvato, Jean Marie Sweigart, Leeann Wille

Photo Credits
©Willem Bosman/Shutterstock, p. vii *bottom;* ©Tim Flach/Getty Images, cover; Getty Images, cover, *bottom left;* ©iStockphoto, pp. iii, v; Royalty-free/Corbis, p. vii *top.*

www.WrightGroup.com

Wright Group

Printed in the United States of America.

Send all inquiries to:
Wright Group/McGraw-Hill
P.O. Box 812960
Chicago, IL 60681

ISBN 0-07-604568-4

7 8 9 CPC 12 11 10 09 08 07

The *McGraw·Hill* Companies

Contents

UNIT 7 | Multiplication and Division

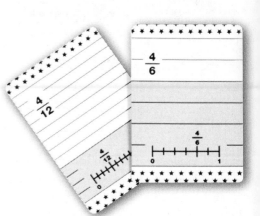

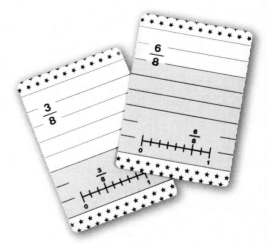

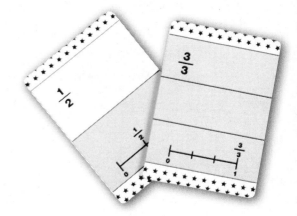

UNIT 10 Measurement and Data

UNIT 11 Probability; Year-Long Projects, Revisited

Activity Sheets

LESSON 7·1 Product Patterns

Part A

Math Message

Complete the facts.

1. $1 \times 1 =$ _____

2. $2 \times 2 =$ _____

3. $3 \times 3 =$ _____

4. $4 \times 4 =$ _____

5. $5 \times 5 =$ _____

6. $6 \times 6 =$ _____

7. $7 \times 7 =$ _____

8. $8 \times 8 =$ _____

9. $9 \times 9 =$ _____

10. $10 \times 10 =$ _____

Part B

A Two's Product Pattern

Multiply. Look for patterns.

11. $2 \times 2 =$ _____

12. $2 \times 2 \times 2 =$ _____

13. $2 \times 2 \times 2 \times 2 =$ _____

14. $2 \times 2 \times 2 \times 2 \times 2 =$ _____

15. $2 \times 2 \times 2 \times 2 \times 2 \times 2 =$ _____

Try This

Use the Two's Product Pattern for Problems 11 through 15. Multiply.

16. $2 \times 2 \times 2 \times 2 \times 2 \times 2 \times 2 =$ _____

LESSON 7·1
Math Boxes

1. This is a picture of a triangular pyramid. This shape has

   _____ faces

   _____ edges

   _____ vertices

   SRB 116

2. Draw an array with 25 Xs arranged in 5 rows.

   How many Xs in each row? _____

   Write a number model for the array.

   _____

   SRB 64 65

3. Draw and label three parallel line segments. Draw and label a line that intersects all three line segments.

   SRB 99 100

4. Fill in the circle next to the correct answer.

   777
   + 1,028
   ‾‾‾‾‾‾

   (A) 251   (B) 1,751

   (C) 1,795   (D) 1,805

   SRB 57–59

5. Complete the Fact Triangle. Write the fact family.

   36

   ×, ÷

   6   _____

   _____ × _____ = _____

   _____ ÷ _____ = _____

   SRB 55

6. Divide the rectangle into 4 equal parts.

LESSON 7·2 Multiplication/Division Facts Table

×,÷	1	2	3	4	5	6	7	8	9	10
1	1	2	3	4	5	6	7	8	9	10
2	2	4	6	8	10	12	14	16	18	20
3	3	6	9	12	15	18	21	24	27	30
4	4	8	12	16	20	24	28	32	36	40
5	5	10	15	20	25	30	35	40	45	50
6	6	12	18	24	30	36	42	48	54	60
7	7	14	21	28	35	42	49	56	63	70
8	8	16	24	32	40	48	56	64	72	80
9	9	18	27	36	45	54	63	72	81	90
10	10	20	30	40	50	60	70	80	90	100

LESSON 7·2   **Math Boxes**

1. Draw the lines of symmetry.

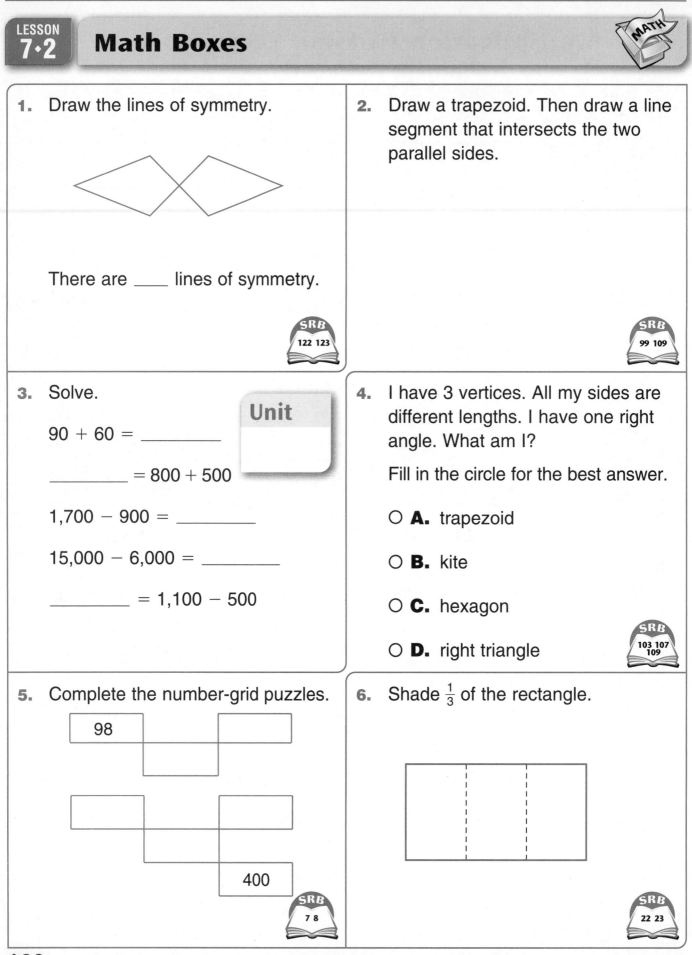

There are ____ lines of symmetry.

SRB
122 123

2. Draw a trapezoid. Then draw a line segment that intersects the two parallel sides.

SRB
99 109

3. Solve.

90 + 60 = _____

_____ = 800 + 500

1,700 − 900 = _____

15,000 − 6,000 = _____

_____ = 1,100 − 500

Unit

4. I have 3 vertices. All my sides are different lengths. I have one right angle. What am I?

Fill in the circle for the best answer.

○ **A.** trapezoid

○ **B.** kite

○ **C.** hexagon

○ **D.** right triangle

SRB
103 107 109

5. Complete the number-grid puzzles.

98

400

SRB
7 8

6. Shade $\frac{1}{3}$ of the rectangle.

SRB
22 23

LESSON 7·3 Multiplication Bingo

Read the rules for *Multiplication Bingo* on pages 293 and 294 in the *Student Reference Book.*

Write the list of numbers on each grid below.

List of numbers

1	9	18	30
4	12	20	36
6	15	24	50
8	16	25	100

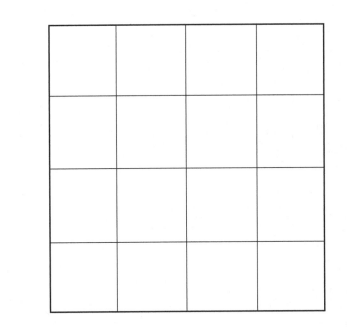

Record the facts you miss.
Be sure to practice them!

_____

_____

_____

_____

_____

Multiplication/Division Practice

Fill in the missing number in each Fact Triangle.
Then write the fact family for the triangle.

1.

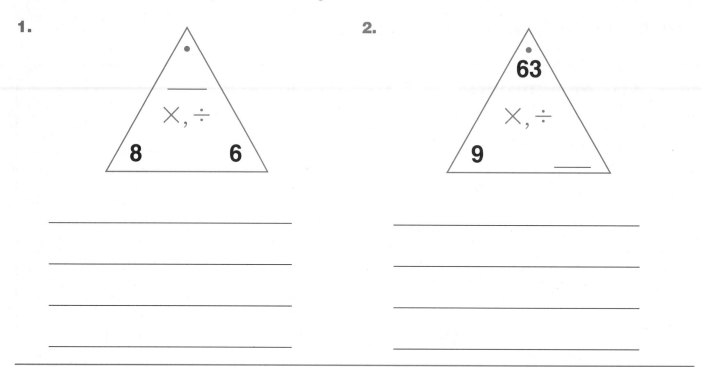

```
          •
       _____
       ×, ÷
    8        6
```

_____

_____

_____

_____

2.

```
          •
        63
       ×, ÷
    9        _____
```

_____

_____

_____

_____

Complete each puzzle.
Example:

×, ÷	3	5
4	*12*	*20*
6	*18*	*30*

3.

×, ÷	2	6
3		
6		

4.

×, ÷	3	5
2		
8		

5.

×, ÷	7	9
2		
5		

6.

×, ÷		4
3	9	
4		

7.

×, ÷		6
2		
	24	36

LESSON 7·3   Math Boxes

1. This is a picture of a cube. What do you know about this shape?

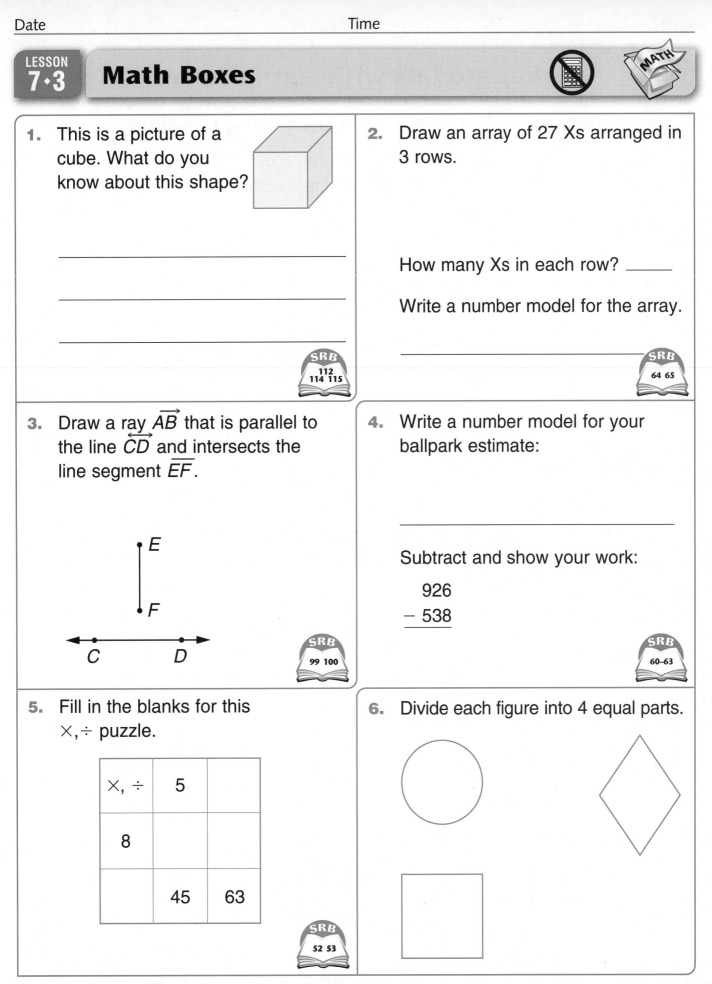

_____

_____

_____

SRB
112
114 115

2. Draw an array of 27 Xs arranged in 3 rows.

How many Xs in each row? _____

Write a number model for the array.

_____

SRB
64 65

3. Draw a ray $\overrightarrow{AB}$ that is parallel to the line $\overleftrightarrow{CD}$ and intersects the line segment $\overline{EF}$.

E

F

C   D

SRB
99 100

4. Write a number model for your ballpark estimate:

_____

Subtract and show your work:

    926
 − 538
 _____

SRB
60–63

5. Fill in the blanks for this ×, ÷ puzzle.

×, ÷	5	
8		
	45	63

SRB
52 53

6. Divide each figure into 4 equal parts.

LESSON 7·4 Number Models with Parentheses

Solve the number story. Then write a number model using parentheses.

1. Alexis scored 12 points, and Nehemie scored 6 points. If their team scored 41 points, how many points did the rest of the team score?

_____

Number model: _____

2. In a partner game, Quincy has 10 points, and Ellen has 14 points. They need 50 points to finish the game. How many more points are needed?

_____

Number model: _____

3. Quincy and Ellen earned 49 points but lost 14 points for a wrong move. They gained 10 points back. What was their score at the end of the round?

_____

Number model: _____

Complete these number sentences.

4. _____ $= 18 - (9 + 5)$

5. $(75 - 29) + 5 =$ _____

6. _____ $= 8 + (9 \times 3)$

7. $36 + (15 \div 3) =$ _____

Add parentheses to complete the number models.

8. $20 - 10 + 4 = 6$

9. $20 - 10 + 4 = 14$

10. $100 - 21 + 10 = 69$

11. $100 - 21 + 10 = 89$

12. $27 - 8 + 3 = 22$

13. $18 = 6 + 3 \times 4$

Math Boxes

1. Draw the lines of symmetry.

There are _____ lines of symmetry.

SRB 122 123

2. Draw a parallelogram. Label the vertices so $\overline{AB} \parallel \overline{CD}$. The symbol $\parallel$ means *is parallel to.*

SRB 108 109

3. Solve.

Unit

_____ = 400 + 800

$3{,}000 + 7{,}000 =$ _____

$90{,}000 - 20{,}000 =$ _____

4. Answer this riddle.

I have four sides. My opposite sides are equal in length. I have two pairs of parallel sides. I do not have any right angles.

What shape am I?

_____

SRB 108 109

5. Complete the number-grid puzzle.

8,742

SRB 7 8

6. Divide the triangles into three equal groups.

△ △ △ △

△ △ △ △

△ △ △ △

LESSON 7·5 Scoring 10 Basketball Points

Find different ways to score 10 points in a basketball game.

Number of 3-point Baskets	Number of 2-point Baskets	Number of 1-point Baskets	Number Models
2	2	0	$(2 \times 3) + (2 \times 2) + (0 \times 1) = 10$

Names with Parentheses

Cross out the names that don't belong in each name-collection box.

1.

12
$(3 \times 3) + 3$ $3 \times (3 + 3)$
$2 + (4 \times 2)$ $(2 + 4) \times 2$
$4 \times (4 - 4)$ $(4 \times 4) - 4$

2.

20
$2 \times (9 + 1)$ $(2 \times 9) + 1$
$30 - (5 \times 2)$ $(30 - 5) \times 2$
$(100 \div 10) + 10$ $100 \div (10 + 10)$

Write names that contain parentheses in each name-collection box.

3.

16

4.

24

5. Write a parentheses problem. Describe how you solved the problem.

_____

_____

_____

_____

LESSON 7·5 | **Math Boxes**

1. Solve.

$(6 \times 3) + 2 =$ _____

$29 - (20 + 3) =$ _____

_____ $= 14 + (3 + 3)$

_____ $= (5 \times 5) - 6$

SRB
16 17

2. Estimate. A bottle of milk costs $2.85. If Nan has $9, does she have enough money to buy 3 bottles of milk? (There is no tax.)

_____

Number model: _____

SRB
191

3. Fill in the oval next to the number model that best describes this array:

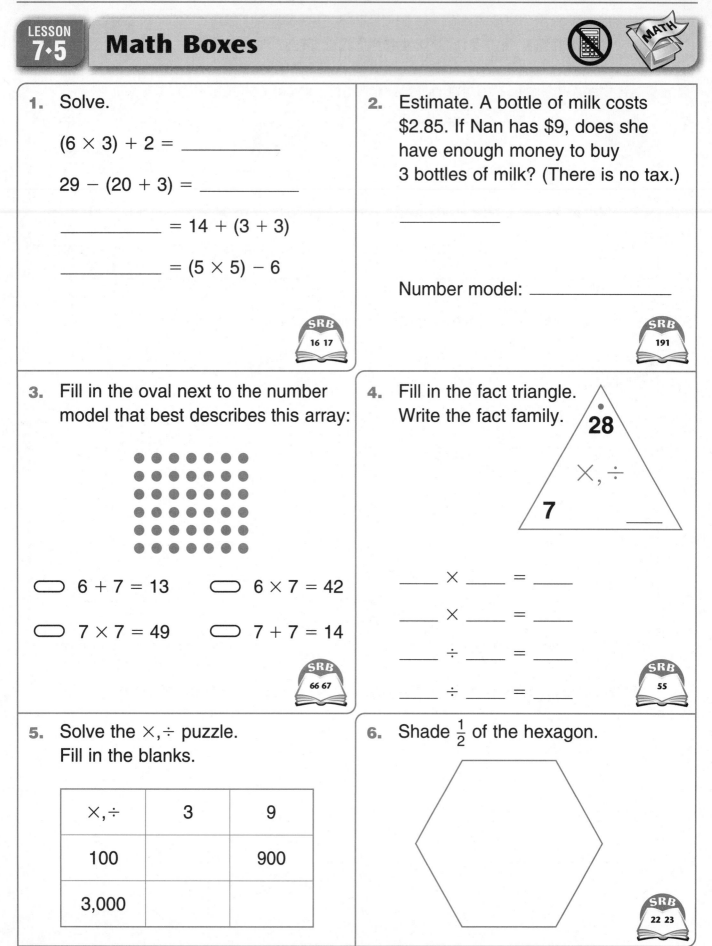

⬭ $6 + 7 = 13$ ⬭ $6 \times 7 = 42$

⬭ $7 \times 7 = 49$ ⬭ $7 + 7 = 14$

SRB
66 67

4. Fill in the fact triangle. Write the fact family.

28

$\times, \div$

7 _____

___ $\times$ ___ $=$ ___

___ $\times$ ___ $=$ ___

___ $\div$ ___ $=$ ___

___ $\div$ ___ $=$ ___

SRB
55

5. Solve the $\times, \div$ puzzle. Fill in the blanks.

$\times, \div$	3	9
100		900
3,000		

6. Shade $\frac{1}{2}$ of the hexagon.

SRB
22 23

168 one hundred sixty-eight

LESSON 7·6

Extended Multiplication and Division Facts

Write the number of 3s in each number.

1. How many 3s in 30? _____

2. How many 3s in 300? _____

3. How many 3s in 3,000? _____

4. How many 3s in 12? _____

5. How many 3s in 120? _____

6. How many 3s in 1,200? _____

Solve each ×,÷ puzzle. Fill in the blanks.

Example:

×,÷	300	2,000
2	600	4,000
3	900	6,000

7.

×,÷	60	300
4		
5	300	

Try This

8.

×,÷	4	5
200		
8,000		

9.

×, ÷		1,000
3	1,500	
		6,000

10. Solve the number story.

A 30-**minute** television program has

◆ two 60-**second** commercials at the beginning,

◆ two 60-**second** commercials at the end, and

◆ four 30-**second** commercials in the middle.

a. How many **minutes** of commercials are there? _____
   (unit)

b. How many **minutes** is the actual program? _____
   (unit)

c. Number model: _____

one hundred sixty-nine **169**

LESSON 7·6

Math Boxes

1. Solve.

   $6 \times 6 =$ _____

   $7 \times 7 =$ _____

   $8 \times 8 =$ _____

   $81 =$ _____ $\times$ _____

   $100 =$ _____ $\times$ _____

   SRB 52 53

2. Fill in the missing whole number factors.

   _____ $\times$ _____ $= 14$

   $28 =$ _____ $\times$ _____

   $32 =$ _____ $\times$ _____

   _____ $\times$ _____ $= 48$

   $54 =$ _____ $\times$ _____

   SRB 52 53

3. Add parentheses to complete the number models.

   $30 = 10 \times 2 + 10$

   $46 - 23 - 13 = 10$

   $4 \div 2 + 6 = 8$

   SRB 16 17

4. Complete.

   in ↓

   Rule

   $\div 2$

   out

in	out
8	
16	
	10
50	

   SRB 203 204

5. Solve.

   $6 \times 10 =$ _____

   $6 \times 30 =$ _____

   $50 \times 6 =$ _____

   _____ $= 70 \times 6$

   $6 \times 90 =$ _____

6. Color $\frac{1}{2}$ of the circle.

   How many fourths are shaded?

   _____ fourths

   SRB 22 23

LESSON 7·7 Stock-up Sale Record

Use the items on pages 216 and 217 in your *Student Reference Book.*

Round 1:

Item to be purchased: _____

How many? _____

Regular or sale price? _____

Price per item: _____

Estimated cost: _____

Round 2:

Item to be purchased: _____

How many? _____

Regular or sale price? _____

Price per item: _____

Estimated cost: _____

Round 3:

Item to be purchased: _____

How many? _____

Regular or sale price? _____

Price per item: _____

Estimated cost: _____

Round 4:

Item to be purchased: _____

How many? _____

Regular or sale price? _____

Price per item: _____

Estimated cost: _____

Round 5:

Item to be purchased: _____

How many? _____

Regular or sale price? _____

Price per item: _____

Estimated cost: _____

Round 6:

Item to be purchased: _____

How many? _____

Regular or sale price? _____

Price per item: _____

Estimated cost: _____

LESSON 7·7 Math Boxes

1. Complete the number models.

$(49 - 19) - 8 = $ _____

$(56 - 14) \times 2 = $ _____

$48 - (19 - 8) = $ _____

$56 - (14 - 2) = $ _____

SRB 16 17

2. Estimate: About how many dollars will Stephen need to buy 4 stopwatches for $12.89 each? (There is no tax.)

Number model:

_____

He will need about

$ _____

SRB 191

3. Solve. Write a number model.

groups	children per group	children in all
6	7	

Number model

_____

SRB 259 260

4. Complete the extended fact triangle. Write the fact family.

_____ $\times$ _____ = _____

_____ $\times$ _____ = _____

_____ $\div$ _____ = _____

_____ $\div$ _____ = _____

SRB 55

5. Solve the $\times, \div$ puzzle. Fill in the blanks.

$\times, \div$		2,000
4	1,200	
		10,000

6. Shade $\frac{1}{2}$ of the balloons.

SRB 22–24

LESSON 7·8 Tens Times Tens

Math Message
Write the dollar values.

1. 10 $10 = $_____

2. 100 $10 = $_____

3. 1,000 $10 = $_____

4. 10 $100 = $_____

5. 100 $100 = $_____

6. 1,000 $100 = $_____

Solve each ×, ÷ puzzle. Fill in the blanks.

7.

×, ÷	10	100
1		
10		

8.

×, ÷	4	30
20		
6		

Try This

9.

×, ÷	40	60
20		
80		

10.

×, ÷		
3	150	
70		560

Multiply.

11. 5 × 90 = _____

12. _____ = 70 × 4

13. 7 × _____ = 420

14. _____ × 90 = 540

15. 10 × 70 = _____

16. 80 × 60 = _____

17. _____ = 30 × 50

18. _____ × 600 = 6,000

LESSON 7·8 Math Boxes

1. Solve.

$49 \div 7 =$ _____

$81 \div 9 =$ _____

_____ $= 64 \div 8$

$6 = 36 \div$ _____

_____ $\div 5 = 5$

SRB 52 53

2. Find the missing factors.

$36 =$ _____ $\times$ _____

$56 =$ _____ $\times$ _____

_____ $\times$ _____ $= 24$

_____ $\times$ _____ $= 42$

$18 =$ _____ $\times$ _____

SRB 52 53

3. Find the missing numbers, and add parentheses to make the number models true.

$3 \times$ _____ $+ 5 = 29$

$5 \times 3 +$ _____ $= 19$

$25 =$ _____ $\times 7 - 3$

SRB 16 17

4. Complete.

in ↓

Rule

$\div 12$

out

inches	feet
12	1
36	
	4
	2
60	

SRB 203 204

5. Solve.

$\begin{array}{r} 8 \\ \times 5 \\ \hline \end{array}$ $\begin{array}{r} 80 \\ \times 5 \\ \hline \end{array}$ $\begin{array}{r} 800 \\ \times 5 \\ \hline \end{array}$

$\begin{array}{r} 8,000 \\ \times 5 \\ \hline \end{array}$ $\begin{array}{r} 5,000 \\ \times 8 \\ \hline \end{array}$

6. Fill in the oval in front of the fraction that does NOT represent the picture.

◯ $\frac{6}{12}$

◯ $\frac{3}{8}$

◯ $\frac{1}{2}$

◯ $\frac{2}{4}$

SRB 27 28

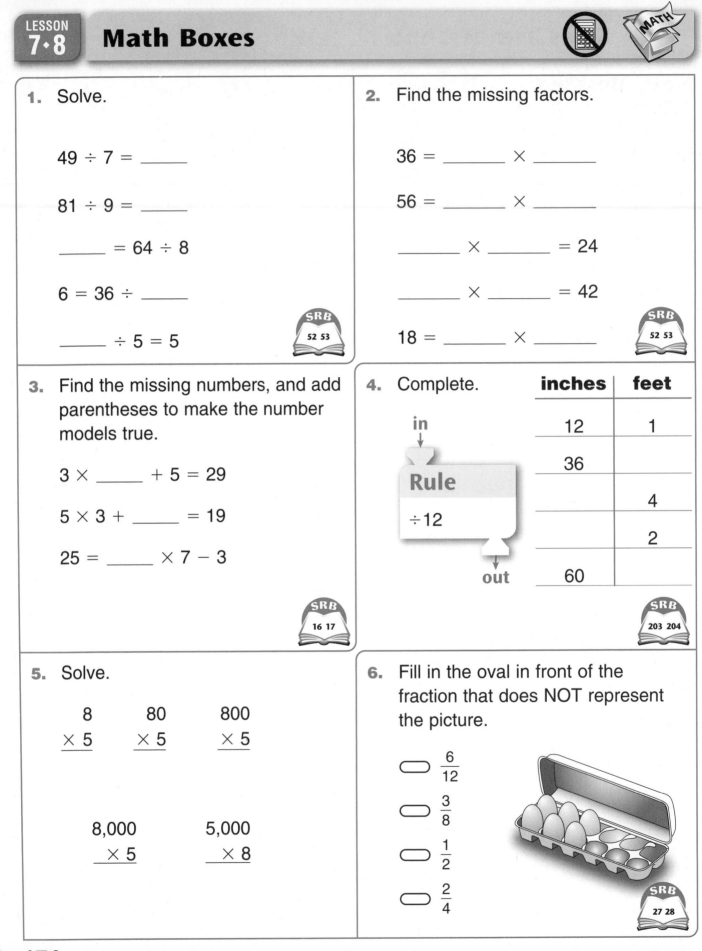

174 one hundred seventy-four

LESSON 7·8

National High/Low Temperatures Project

Date	Highest Temperature (maximum)		Lowest Temperature (minimum)		Difference (range)
	Place	Temperature	Place	Temperature	
		°F		°F	°F
		°F		°F	°F
		°F		°F	°F
		°F		°F	°F
		°F		°F	°F
		°F		°F	°F
		°F		°F	°F
		°F		°F	°F
		°F		°F	°F
		°F		°F	°F
		°F		°F	°F
		°F		°F	°F
		°F		°F	°F
		°F		°F	°F
		°F		°F	°F
		°F		°F	°F
		°F		°F	°F
		°F		°F	°F
		°F		°F	°F
		°F		°F	°F
		°F		°F	°F

LESSON 7·8 Temperature Ranges Graph

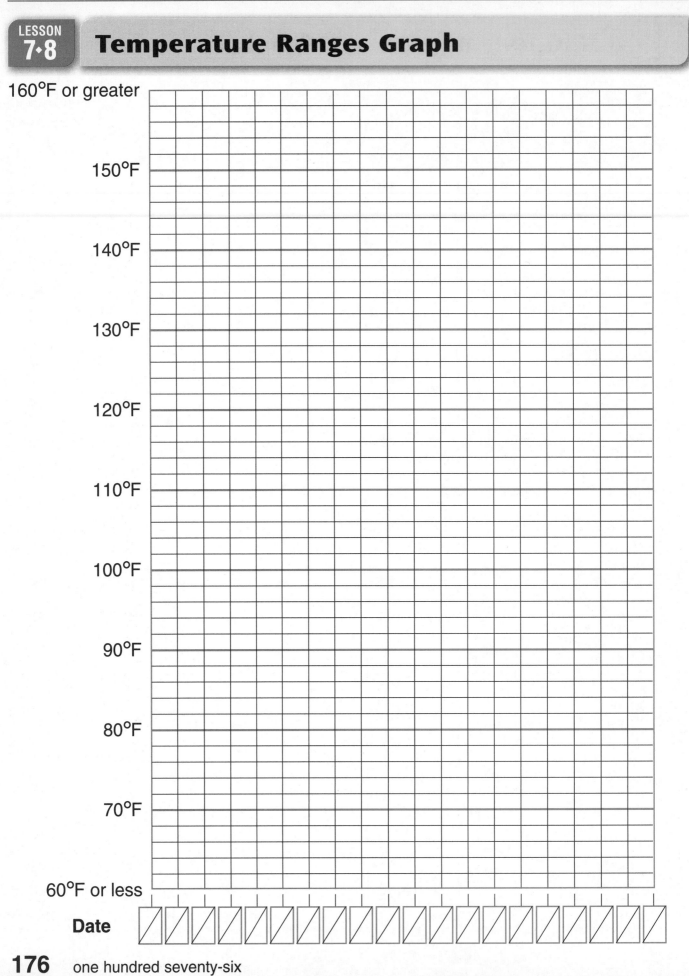

LESSON 7·8 Temperature Ranges Graph *continued*

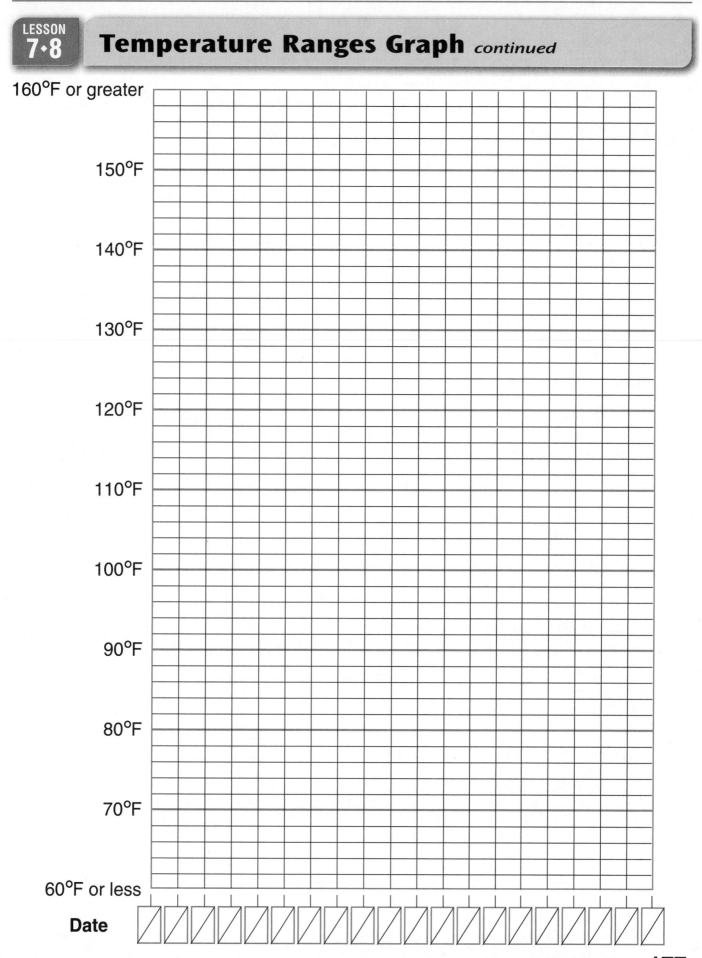

160°F or greater

150°F

140°F

130°F

120°F

110°F

100°F

90°F

80°F

70°F

60°F or less

Date

LESSON 7·9   **Math Boxes**

1. Add parentheses to complete the number models.

$14 - 7 \times 2 = 14$

$3 \times 6 + 2 = 24$

$7 = 6 + 15 \div 3$

$9 \times 5 + 3 = 72$

SRB 16 17

2. Write a number model for your ballpark estimate:

_____

Subtract. Show your work.

$\begin{array}{r} 900 \\ -\ 799 \\ \hline \end{array}$

SRB 192

3. Solve. Show your work.

7 cartons
6 donuts per carton

How many donuts in all?

_____ donuts

SRB 259 260

4. Complete the extended Fact Triangle. Write the fact family.

_____

_____

_____

_____

2,400

$\times, \div$

8   _____

5. Three of the names do not belong in this name-collection box. Cross them out.

4,000

$8 \times 5{,}000$ $(500 \times 5) - 500$

$5{,}000 - (5 \times 200)$

$(200 \times 4) \times 5$ $2 \times 2{,}000$

$(200 \div 4) \times 8$ $1{,}000 \times 4$

$8{,}000 \div 2$ $(2 \times 2) \times 1{,}000$

$(200 + 200) \times 10$

6. Divide the triangles into 2 equal groups.

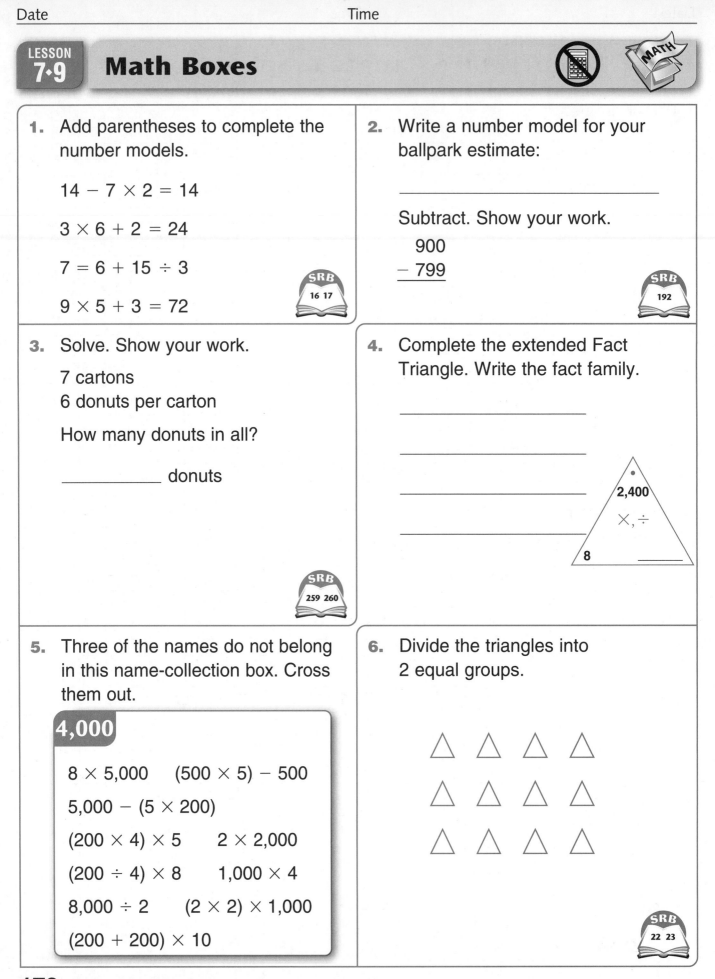

SRB 22 23

LESSON 7·10   Math Boxes

1. Circle the pictures in which $\frac{1}{2}$ is shaded.

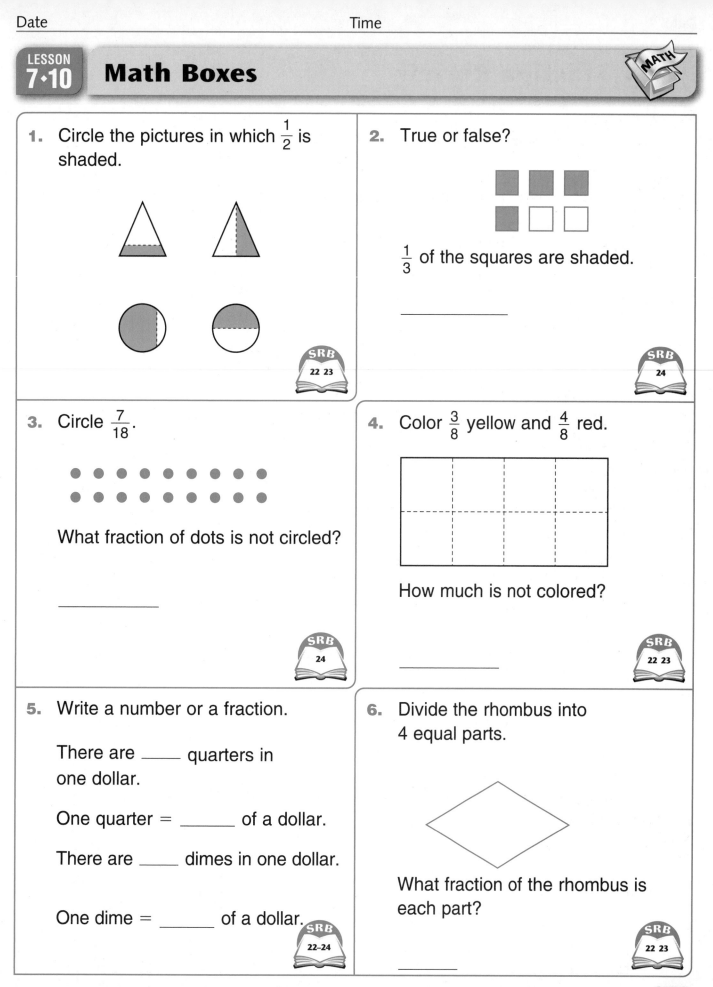

SRB 22 23

2. True or false?

$\frac{1}{3}$ of the squares are shaded.

_____

SRB 24

3. Circle $\frac{7}{18}$.

What fraction of dots is not circled?

_____

SRB 24

4. Color $\frac{3}{8}$ yellow and $\frac{4}{8}$ red.

How much is not colored?

_____

SRB 22 23

5. Write a number or a fraction.

There are _____ quarters in one dollar.

One quarter = _____ of a dollar.

There are _____ dimes in one dollar.

One dime = _____ of a dollar.

SRB 22–24

6. Divide the rhombus into 4 equal parts.

What fraction of the rhombus is each part?

_____

SRB 22 23

LESSON 8·1 Fraction Review

Math Message

1. Draw an X through $\frac{2}{3}$ of the circles. ◯ ◯ ◯ ◯ ◯ ◯

Label each picture with one of the following numbers: $0, \frac{0}{4}, \frac{1}{4}, \frac{1}{2}, \frac{2}{4},$ or $\frac{3}{4}$.

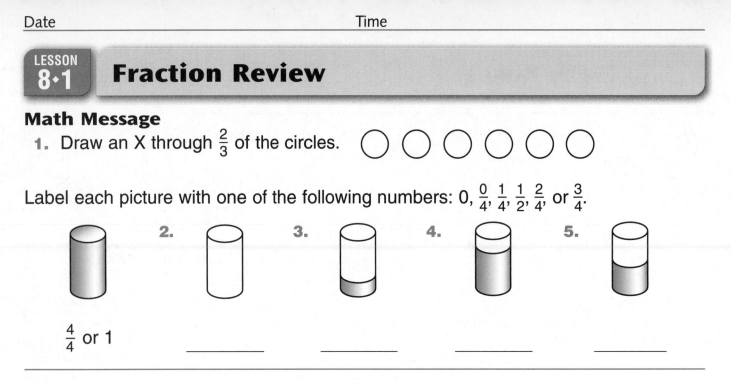

2.   **3.**   **4.**   **5.**

$\frac{4}{4}$ or 1   _____   _____   _____   _____

Each whole figure represents ONE.
Write a fraction that names each region inside the figure.

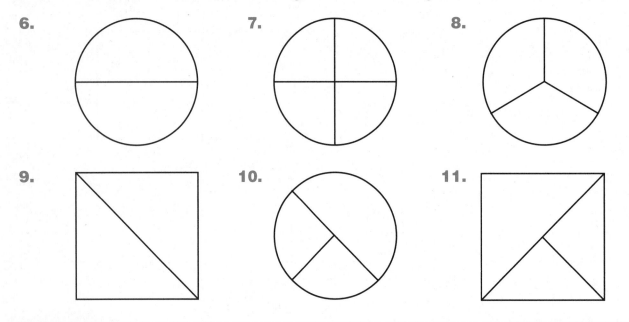

6.   7.   8.

9.   10.   11.

Try This

Each whole figure represents ONE. Write a fraction that names each
region inside the figure.

12.

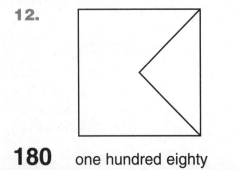

13.

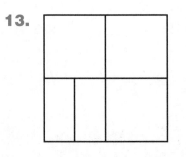

LESSON 8·1   Fraction Review *continued*

You need at least 25 pennies or other counters to help you solve these problems. Share solution strategies with others in your group.

> **Unit**
> counters

14. a. Show $\frac{1}{4}$ of a set of 8 counters. How many counters is that? _____

b. Show $\frac{2}{4}$ of the set. How many counters? _____

c. Show $\frac{3}{4}$ of the set. How many counters? _____

15. a. Show $\frac{1}{3}$ of a set of 12 counters. How many counters is that? _____

b. Show $\frac{2}{3}$ of the set. How many counters? _____

c. Show $\frac{3}{3}$ of the set. How many counters? _____

16. a. Show $\frac{1}{5}$ of a set of 15 counters. How many counters is that? _____

b. Show $\frac{4}{5}$ of the set. How many counters? _____

17. Show $\frac{3}{4}$ of a set of 20 counters. How many counters? _____

18. Show $\frac{2}{3}$ of a set of 18 counters. How many counters? _____

19. Five counters is $\frac{1}{5}$ of a set. How many counters are in the whole set? _____

20. Six counters is $\frac{1}{3}$ of a set. How many counters are in the whole set? _____

> **Try This**

21. Twelve counters is $\frac{3}{4}$ of a set. How many counters are in the complete set? _____

22. Pretend that you have a set of 15 cheese cubes. What is $\frac{1}{2}$ of that set? Use a fraction or decimal in your answer. _____

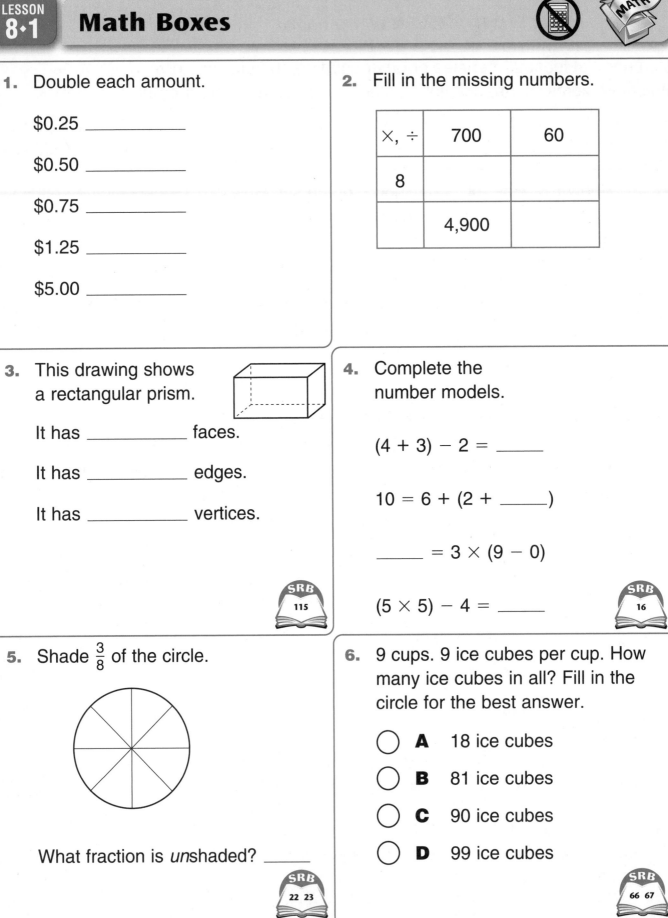

LESSON 8·1 Math Boxes

1. Double each amount.

$0.25 _____

$0.50 _____

$0.75 _____

$1.25 _____

$5.00 _____

2. Fill in the missing numbers.

×, ÷	700	60
8		
	4,900	

3. This drawing shows a rectangular prism.

It has _____ faces.

It has _____ edges.

It has _____ vertices.

SRB 115

4. Complete the number models.

(4 + 3) − 2 = _____

10 = 6 + (2 + _____)

_____ = 3 × (9 − 0)

(5 × 5) − 4 = _____

SRB 16

5. Shade $\frac{3}{8}$ of the circle.

What fraction is *un*shaded? _____

SRB 22 23

6. 9 cups. 9 ice cubes per cup. How many ice cubes in all? Fill in the circle for the best answer.

○ **A** 18 ice cubes

○ **B** 81 ice cubes

○ **C** 90 ice cubes

○ **D** 99 ice cubes

SRB 66 67

LESSON 8·2 **Drawing Blocks**

Color the blocks in the bags blue. Then fill in the blanks by answering this question: How many red blocks would you put into each bag?

1. If I wanted to be sure of taking
   out a blue block, I would put in _____ red block(s).

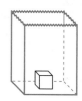

2. If I wanted to have an equal chance
   of taking out red or blue, I would put in _____ red block(s).

3. If I wanted to be more likely to take
   out blue than red, I would put in _____ red block(s).

4. If I wanted to take out a red block about
   3 times as often as a blue one, I would put in _____ red block(s).

5. If I wanted to take out a red block about
   half as often as a blue one, I would put in _____ red block(s).

6. If I wanted to take out a red block
   about $\frac{1}{3}$ of the time, I would put in _____ red block(s).

Try This

7. If I wanted to take out a red block
   about $\frac{2}{3}$ of the time, I would put in _____ red block(s).

LESSON 8·2 — Math Boxes

1. Shade $\frac{3}{7}$ of the books.

SRB
24

2. Complete the bar graph.

Max swam 5 laps.

Colin swam 3 laps.

Miles swam 6 laps.

Median number of laps:

_____

Laps Swum

Max   Colin   Miles

SRB
80 86

3. Circle $\frac{5}{10}$ of the collection of triangles.

Write 2 names for the fraction that is left.

_____ and _____

SRB
24

4. What is the missing factor? Fill in the circle for the best answer.

$6 \times$ _____ $= 3,600$

(A) 6

(B) 60

(C) 600

(D) 6,000

5. Flip 1 penny and 1 nickel. Show all possible outcomes. Use Ⓟ and Ⓝ to show the coins. Use H for HEADS and T for TAILS.

6. How much do six 300-pound dolphins weigh?

dolphins	pounds per dolphin	pounds in all

Answer: _____

Number model:

_____

SRB
259 260

LESSON 8·3   Fractions with Pattern Blocks

Exploration A

Work with a partner.

Materials   ☐ pattern blocks
☐ Pattern-Block Template

Part 1

Cover each shape with green △ pattern blocks. What fractional part of each shape is 1 green pattern block? Write the fraction under each shape.

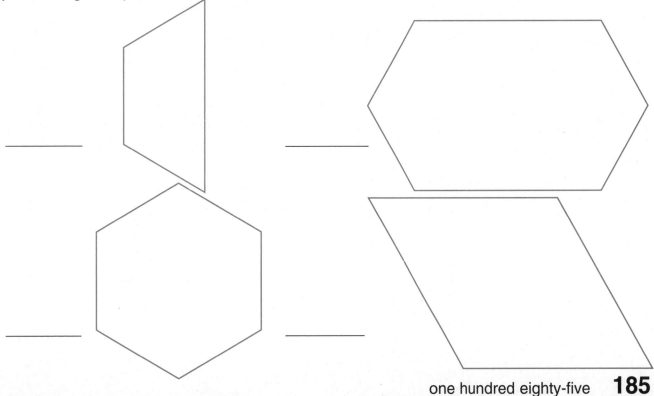

Part 2

Cover each shape with green △ pattern blocks. What fractional part of each shape are 2 green pattern blocks? Write the fraction next to each shape.

LESSON 8·3 **Fractions with Pattern Blocks** *continued*

Part 3

Cover each shape with blue pattern blocks. What fractional part of each shape is 1 blue pattern block? Write the fraction under each shape. If you can't cover the whole shape, cover as much as you can. *Think:* Is there another block that would cover the rest of the shape?

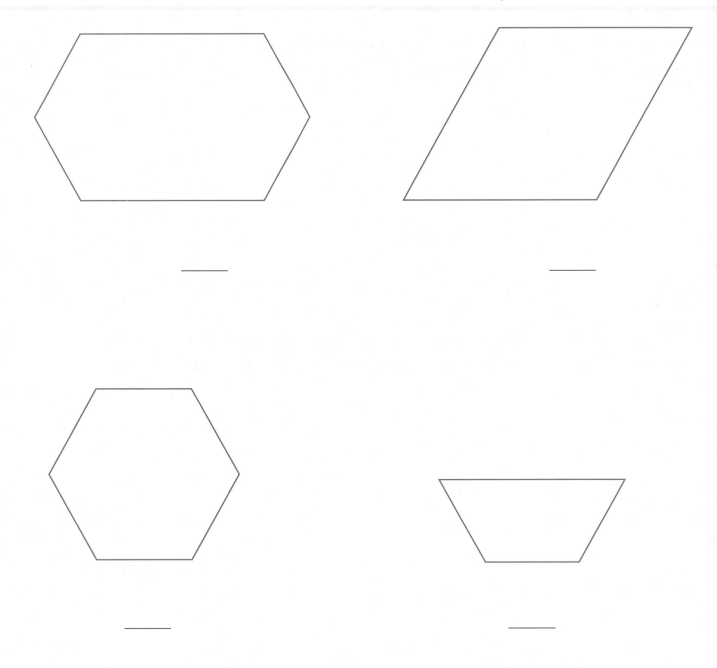

LESSON 8·3 Fractions with Pattern Blocks *continued*

Try This

Part 4

Cover each shape with blue 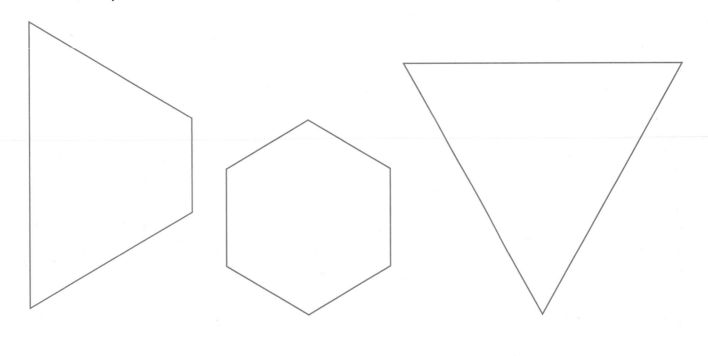 pattern blocks. What fractional part of each shape would 2 blue pattern blocks cover? Write the fraction under each shape.

_____   _____   _____

Part 5

Use your Pattern-Block Template to show how you divided the shapes in each section. *Remember:* The number *under* the fraction bar names the number of equal parts into which the whole shape is divided.

Follow-Up

Get together with the rest of the group.

◆ Compare your answers.

◆ Use the blocks to check your answers.

◆ Decide whether more than one fraction can be correct.

LESSON 8·3 Dressing for the Party

Exploration C

Work in a group of four.

Materials
- ☐ *Math Masters,* p. 244 (Pants and Socks Cutouts)
- ☐ scissors
- ☐ tape
- ☐ blue, red, green, and black crayons or coloring pencils

Problem

Pretend that you have 4 pairs of pants: blue, red, green, and black. You also have 4 pairs of socks: blue, red, green, and black. You have been invited to a party. You need to choose a pair of pants and a pair of socks to wear. (Of course, both socks must be the same color.) For example, the pants could be blue and both socks black.

How many different combinations of pants and socks are possible?

Strategy

Use the cutouts on *Math Masters,* page 244, and crayons to help you answer the question.

Before you answer the question, decide on a way for your group to share the following work.

- ◆ Color the pants in the first row blue.
- ◆ Color the pants in the second row red.
- ◆ Color the pants in the third row green and those in the fourth row black.
- ◆ Color the socks in the same way.
- ◆ Cut out each pair of pants and each pair of socks.
- ◆ Tape together pairs of pants and pairs of socks to show different outfits. Check that you have only one of each outfit.

LESSON 8·3 **Dressing for the Party** *continued*

1. How many different combinations of
   pants and socks did your group find? _____

2. Is this all of the possible combinations? _____

3. How do you know?

   _____

   _____

   _____

   _____

4. How did your group divide up the work?

   _____

   _____

   _____

   _____

5. How did your group solve the problem?

   _____

   _____

   _____

   _____

LESSON 8·3 Math Boxes

1. Double the amounts.

$1.10 _____

$2.50 _____

$10.50 _____

$12.50 _____

$25.00 _____

2. Solve.

$5 \times 9 =$ _____

$5 \times 90 =$ _____

$5 \times 900 =$ _____

_____ $= 3 \times 8$

_____ $= 30 \times 80$

_____ $= 300 \times 80$

3. This drawing shows a square pyramid.

It has _____ faces.

It has _____ edges.

It has _____ vertices.

What is the shape of its base?

_____

SRB 116

4. Put in the parentheses needed to complete the number models.

$31 = 3 + 7 \times 4$

$40 = 3 + 7 \times 4$

$4 \times 8 + 2 \times 2 = 36$

$4 \times 8 + 2 \times 2 = 80$

SRB 16 17

5. Color $\frac{2}{5}$ of the rectangle.

What fraction is *not* colored? _____

SRB 22 23

6. 12 apples per bag.

How many apples in 3 bags?

_____

How many apples in 4 bags?

_____

SRB 66 67

LESSON 8·4 Fraction Number-Line Poster

1 Whole
Halves
Fourths
Eighths
Thirds
Sixths

LESSON 8·4 Frames-and-Arrows Problems

Solve each Frames-and-Arrows problem. Use your Fraction Number-Line Poster on *Math Journal 2,* page 191 for Problems 1 and 2.

SRB 200 201

1.

Rule

$\frac{1}{8}$ more

| $\frac{3}{8}$ | $\frac{4}{8}$ | | | $\frac{7}{8}$ | $\frac{8}{8}$ |

2.

Rule

$-\frac{1}{6}$

| $\frac{5}{6}$ | | | | $\frac{1}{6}$ | |

3.

Rule

| 3 | | 12 | | 48 | |

4.

| 10¢ | 35¢ | | 50¢ | 40¢ | 65¢ |

Try This

5.

×5

−50

| 12 | | 10 | | |

LESSON 8·4

Math Boxes

1. Shade $\frac{7}{10}$ of the hats.

SRB
24

2. Use the bar graph.

Maximum:

_____
number of
laps

Minimum:

_____
number of
laps

Range: _____ number of laps

Laps Swum

Max   Colin   Miles

3. Suppose you like pizza and are very hungry. Would you rather have $\frac{4}{5}$ of a pizza or $\frac{8}{10}$ of a pizza?

_____

Why? _____

_____

_____

4. Fill in the missing numbers.

×, ÷		600
50	1,500	
		42,000

5. True or false? There is an equal chance of taking a B or an R block out of the bag.

_____

6. How much do seven packs of pencils cost if each pack costs $0.80?

packs of pencils	cost per pack	cost in all

Answer: _____

Number model:

_____

SRB
259 260

LESSON 8·5 Table of Equivalent Fractions

Use your deck of Fraction Cards to find equivalent fractions.
Record them in the table.

Fraction	Equivalent Fractions
$\frac{0}{2}$	
$\frac{1}{2}$	
$\frac{2}{2}$	
$\frac{1}{3}$	
$\frac{2}{3}$	
$\frac{1}{4}$	
$\frac{3}{4}$	
$\frac{1}{5}$	
$\frac{4}{5}$	
$\frac{1}{6}$	
$\frac{5}{6}$	

Describe any patterns you see.

_____

_____

_____

_____

LESSON 8·5   **Math Boxes**

1. In the number 3.514:

   the 3 means ___*3 ones*___

   the 1 means _____

   the 5 means _____

   the 4 means _____

   SRB 35

2. Which is true? Fill in the circle for the best answer.

   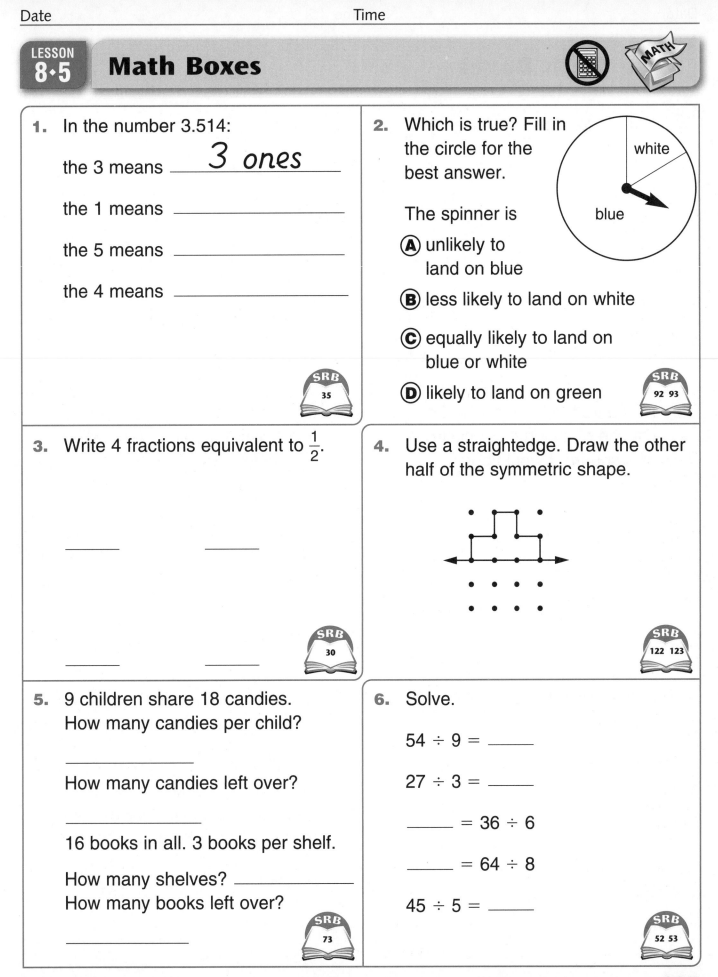

   The spinner is

   Ⓐ unlikely to land on blue

   Ⓑ less likely to land on white

   Ⓒ equally likely to land on blue or white

   Ⓓ likely to land on green

   SRB 92 93

3. Write 4 fractions equivalent to $\frac{1}{2}$.

   _____   _____

   _____   _____

   SRB 30

4. Use a straightedge. Draw the other half of the symmetric shape.

   SRB 122 123

5. 9 children share 18 candies. How many candies per child?

   _____

   How many candies left over?

   _____

   16 books in all. 3 books per shelf.

   How many shelves? _____
   How many books left over?

   _____

   SRB 73

6. Solve.

   $54 \div 9 =$ _____

   $27 \div 3 =$ _____

   _____ $= 36 \div 6$

   _____ $= 64 \div 8$

   $45 \div 5 =$ _____

   SRB 52 53

LESSON 8·6   Math Boxes

1. Write 4 fractions equivalent to $\frac{1}{4}$.

_____   _____

_____   _____

SRB
30

2. Complete.

_____ hours = 1 day

12 hours = _____ day

_____ weeks = 21 days

_____ minutes = $\frac{1}{2}$ hour

15 minutes = _____ hour

SRB
247

3. If I wanted to have an equal chance of taking out a circle or a square, I would put in

_____ circle(s).

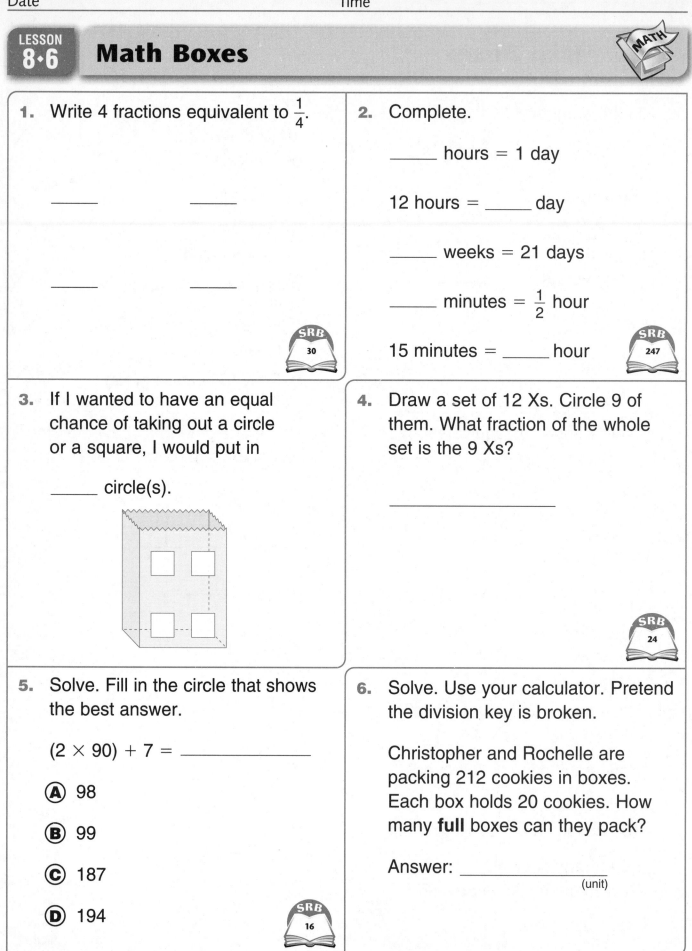

4. Draw a set of 12 Xs. Circle 9 of them. What fraction of the whole set is the 9 Xs?

_____

SRB
24

5. Solve. Fill in the circle that shows the best answer.

$(2 \times 90) + 7 =$ _____

Ⓐ 98

Ⓑ 99

Ⓒ 187

Ⓓ 194

SRB
16

6. Solve. Use your calculator. Pretend the division key is broken.

Christopher and Rochelle are packing 212 cookies in boxes. Each box holds 20 cookies. How many **full** boxes can they pack?

Answer: _____
                              (unit)

196   one hundred ninety-six

LESSON 8·7   More Than ONE

Use the circles that you cut out for the Math Message.

1. Glue 3 halves into the two whole circles.

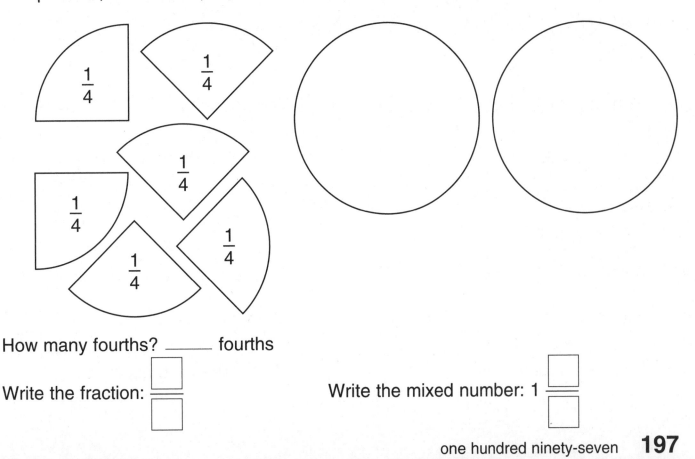

$\frac{1}{2}$ $\frac{1}{2}$ $\frac{1}{2}$

3 halves or $\frac{3}{2}$

$1\frac{1}{2}$ or one and 1 half

2. Glue 6 fourths into the two whole circles. Fill in the missing digits in the question, the fraction, and the mixed number.

$\frac{1}{4}$ $\frac{1}{4}$ $\frac{1}{4}$ $\frac{1}{4}$ $\frac{1}{4}$ $\frac{1}{4}$

How many fourths? _____ fourths

Write the fraction: $\dfrac{\boxed{}}{\boxed{}}$

Write the mixed number: 1 $\dfrac{\boxed{}}{\boxed{}}$

LESSON 8·7   **More Than ONE** *continued*

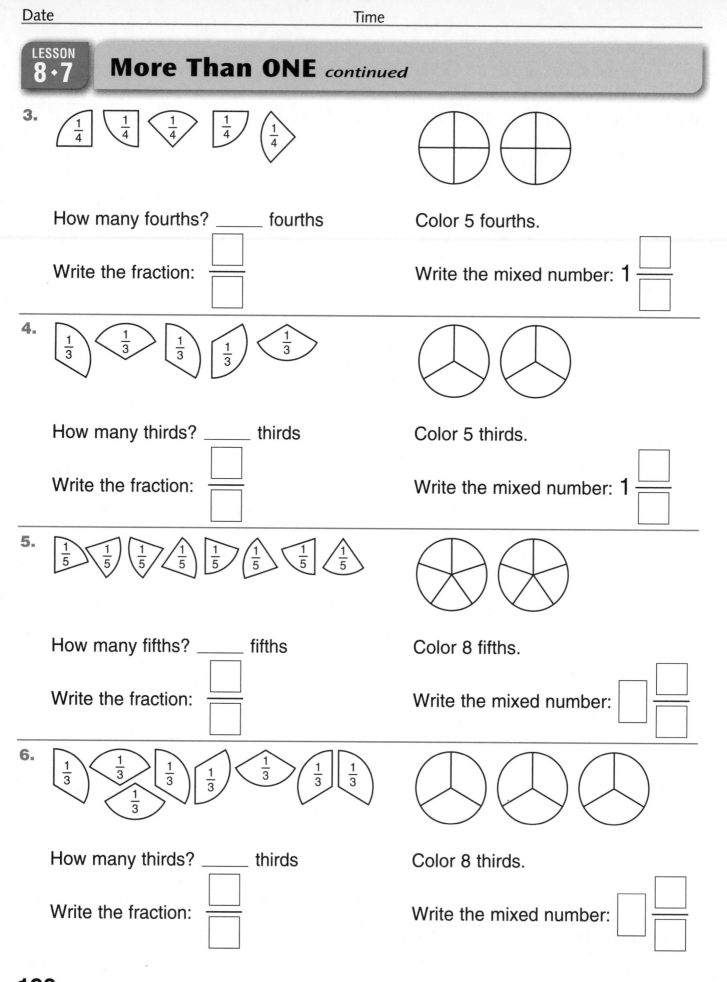

3. How many fourths? _____ fourths

Write the fraction: ☐/☐

Color 5 fourths.

Write the mixed number: 1 ☐/☐

4. How many thirds? _____ thirds

Write the fraction: ☐/☐

Color 5 thirds.

Write the mixed number: 1 ☐/☐

5. How many fifths? _____ fifths

Write the fraction: ☐/☐

Color 8 fifths.

Write the mixed number: ☐ ☐/☐

6. How many thirds? _____ thirds

Write the fraction: ☐/☐

Color 8 thirds.

Write the mixed number: ☐ ☐/☐

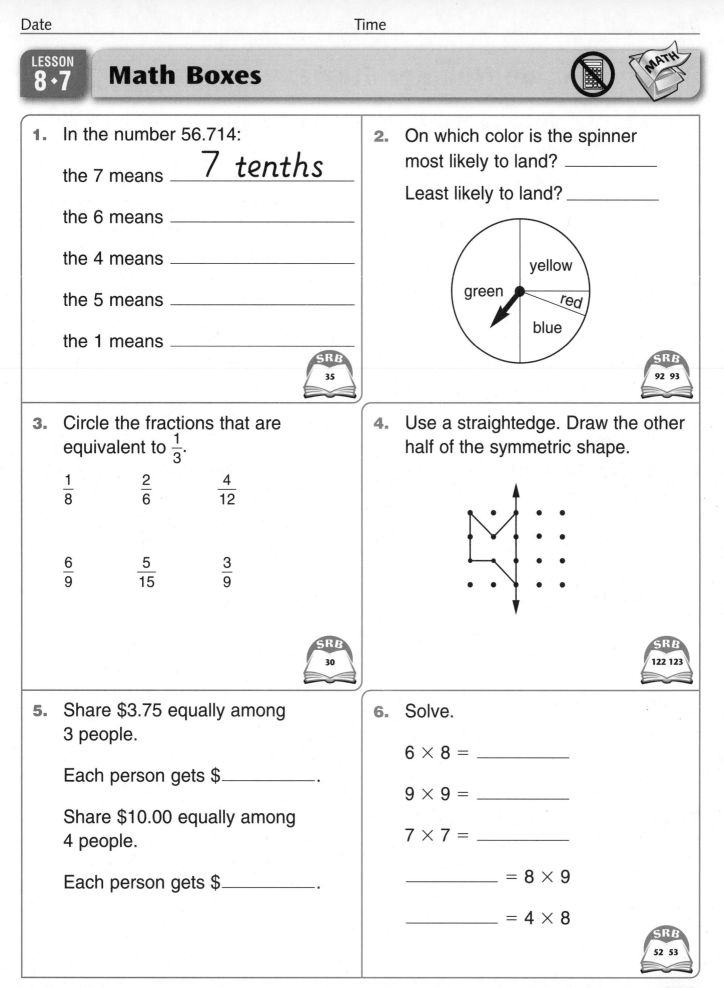

LESSON 8·7

Math Boxes

1. In the number 56.714:

   the 7 means _**7 tenths**_

   the 6 means _____

   the 4 means _____

   the 5 means _____

   the 1 means _____

   SRB 35

2. On which color is the spinner most likely to land? _____

   Least likely to land? _____

   yellow

   green

   red

   blue

   SRB 92 93

3. Circle the fractions that are equivalent to $\frac{1}{3}$.

   $\frac{1}{8}$ $\frac{2}{6}$ $\frac{4}{12}$

   $\frac{6}{9}$ $\frac{5}{15}$ $\frac{3}{9}$

   SRB 30

4. Use a straightedge. Draw the other half of the symmetric shape.

   SRB 122 123

5. Share $3.75 equally among 3 people.

   Each person gets $_____.

   Share $10.00 equally among 4 people.

   Each person gets $_____.

6. Solve.

   $6 \times 8 =$ _____

   $9 \times 9 =$ _____

   $7 \times 7 =$ _____

   _____ $= 8 \times 9$

   _____ $= 4 \times 8$

   SRB 52 53

LESSON 8·8   Fraction Number Stories

Solve these number stories. Use pennies, counters, or draw
pictures to help you.

1. There are 8 apples in the package.
Glenn did not eat any. What fraction
of the package did Glenn eat?

   _____

2. Anik bought a dozen eggs at the
supermarket. When he got home, he
found that $\frac{1}{6}$ of the eggs were cracked.
How many eggs were cracked?

   _____ eggs

3. Chante used $\frac{2}{3}$ of a package of
ribbon to wrap presents. Did she use
more or less than $\frac{3}{4}$ of the package?

   _____

4. I had 2 whole cookies. I gave you $\frac{1}{4}$
of 1 cookie. How many cookies did I
have left?

   _____ cookies

5. There are 10 quarters. You have 3.
I have 2. What fraction of the
quarters do you have?

   _____

   What fraction of the quarters do
I have?

   _____

   What fraction of the quarters do
we have together?

   _____

6. One day, Edwin read $\frac{1}{3}$ of a book.
The next day, he read another $\frac{1}{3}$ of
the book. What fraction of the book
had he read after 2 days?

   _____

   What fraction of the book did he
have left to read?

   _____

7. Dorothy walks $1\frac{1}{2}$ miles to school.
Jaime walks $1\frac{2}{4}$ miles to school.
Who walks the longer distance?

   _____

8. Twelve children shared
2 medium-size pizzas equally.
What fraction of 1 whole pizza
did each child eat?

   _____

LESSON
8·8   **Fraction Number Stories** *continued*

9. Write a fraction story. Ask your partner to solve it.

_____

_____

_____

_____

_____

Draw eggs in each carton to show the fraction.

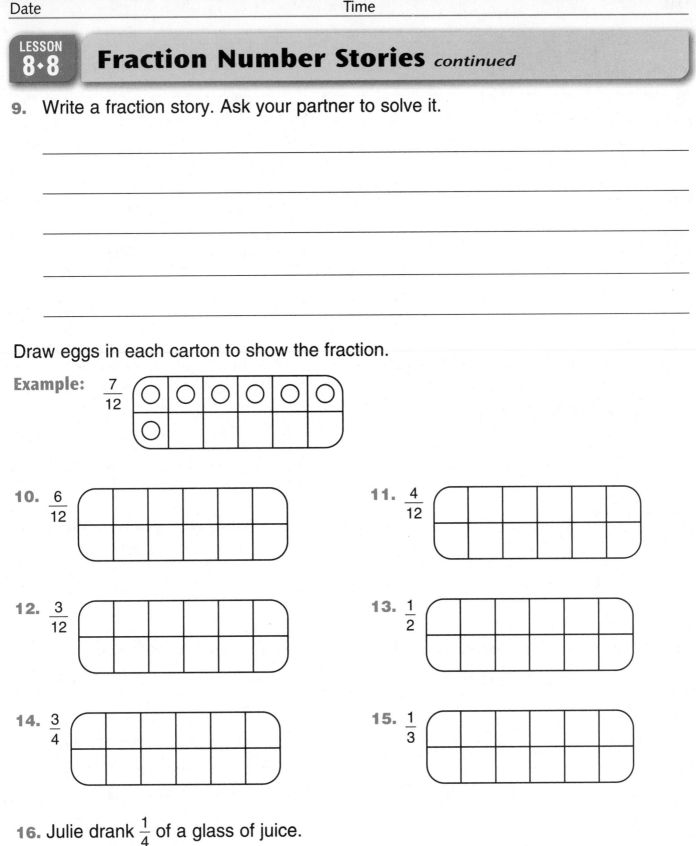

Example: $\frac{7}{12}$

10. $\frac{6}{12}$

11. $\frac{4}{12}$

12. $\frac{3}{12}$

13. $\frac{1}{2}$

14. $\frac{3}{4}$

15. $\frac{1}{3}$

16. Julie drank $\frac{1}{4}$ of a glass of juice.

Draw an empty glass.

Shade in the glass to show how much juice is left. Write the fraction.

____ of the glass of juice is left.

Math Boxes

1. Draw two ways to show $\frac{2}{3}$.

SRB
22–24

2. 6 feet = _____ yards

_____ feet = 18 inches

$1\frac{1}{3}$ yards = _____ feet

$1\frac{1}{2}$ yards = _____ inches

SRB
246

3. Use simple drawings to show all of the possible ways you can take 2 blocks from the bag.

4. Tara frosted $\frac{4}{5}$ of the cupcakes. What fraction of the cupcakes

is *not* frosted? _____

Did she frost more or less than $\frac{1}{2}$ of the cupcakes? _____

If there were 20 cupcakes in all, how many did she frost?

_____

5. Show two ways a team can score 37 points in a football game.

7 points	6 points	3 points	2 points

Write a number model:

_____

_____

SRB
16 17

6. Use your calculator. Pretend the division key is broken. Solve this problem.

Will, Wes, Sam, and Ameer want to share $25 equally. How much money will each person get?

Answer: _____

LESSON 8·9 Math Boxes

1. Solve.

$7 \times 6 =$ _____

$7 \times 60 =$ _____

$7 \times 600 =$ _____

_____ $= 8 \times 6$

_____ $= 8 \times 60$

_____ $= 8 \times 600$

2. Share $2.70 equally among 3 people.

Each person gets $_____.

Share $9 equally among 4 people.

Each person gets $_____.

3. 30 is 10 times as much as

_____.

500 is _____ times as much as 5.

_____ is 100 times as much as 80.

40,000 is 1,000 times as much

as _____.

4. 6 coats. 4 buttons per coat. How many buttons in all?

_____ buttons

Write a number model.

_____

5. Draw a 4-by-7 array of Xs.

How many Xs in all? _____

Write a number model.

_____

6. 18 books. 6 books per shelf.

How many shelves? _____

How many books left over? _____

4 children share 13 marbles. How many marbles per child?

_____

How many marbles left over?

_____

SRB
64 65

LESSON 9·1 Adult Weights of North American Animals

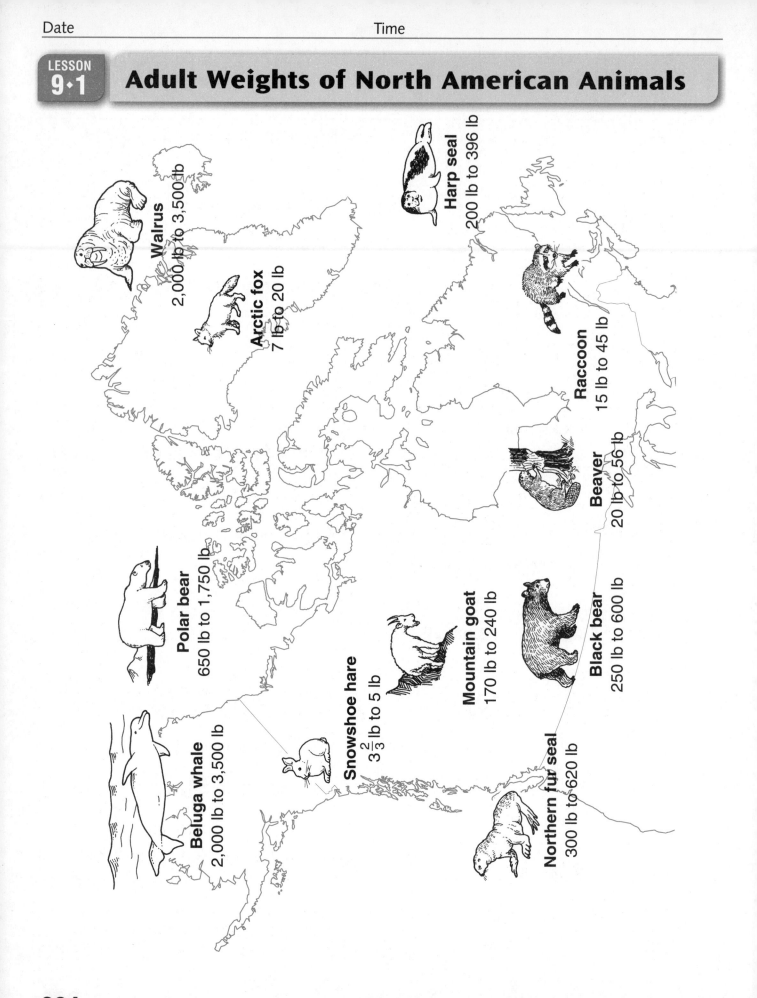

Harp seal
200 lb to 396 lb

Walrus
2,000 lb to 3,500 lb

Arctic fox
7 lb to 20 lb

Raccoon
15 lb to 45 lb

Beaver
20 lb to 56 lb

Polar bear
650 lb to 1,750 lb

Mountain goat
170 lb to 240 lb

Black bear
250 lb to 600 lb

Snowshoe hare
$3\frac{2}{3}$ lb to 5 lb

Beluga whale
2,000 lb to 3,500 lb

Northern fur seal
300 lb to 620 lb

LESSON 9·1

Adult Weights of North American Animals *cont.*

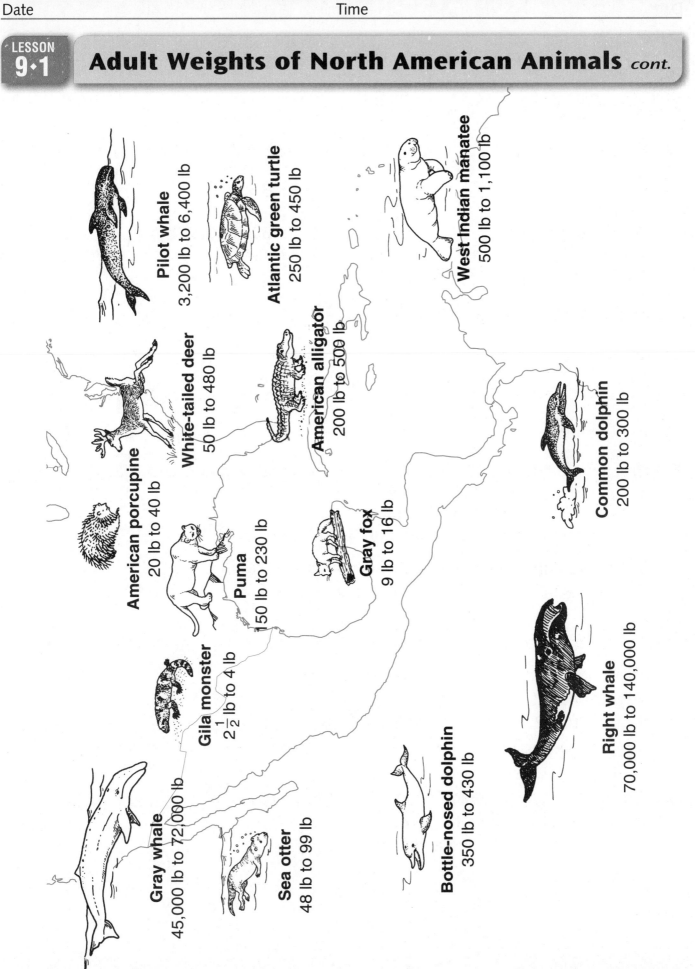

Pilot whale
3,200 lb to 6,400 lb

Atlantic green turtle
250 lb to 450 lb

West Indian manatee
500 lb to 1,100 lb

White-tailed deer
50 lb to 480 lb

American alligator
200 lb to 500 lb

Common dolphin
200 lb to 300 lb

American porcupine
20 lb to 40 lb

Puma
150 lb to 230 lb

Gray fox
9 lb to 16 lb

Gila monster
$2\frac{1}{2}$ lb to 4 lb

Gray whale
45,000 lb to 72,000 lb

Sea otter
48 lb to 99 lb

Bottle-nosed dolphin
350 lb to 430 lb

Right whale
70,000 lb to 140,000 lb

LESSON 9·1   Multiples of 10, 100, and 1,000

Solve each problem.

1. a. 7 [40s] = _____   b. 7 × 40 = _____

2. a. 6 [70s] = _____   b. 6 × 70 = _____

3. a. 60 [20s] = _____   b. 60 × 20 = _____

4. How many 50s are in 4,000? _____

5. How many 800s are in 2,400? _____

6. a. How many 3s are in 270? _____   b. _____ × 3 = 270

   c. 270 ÷ 3 = _____

7. a. 40 × 300 = _____   b. 12,000 ÷ 40 = _____

For Problems 8 through 10, use the information on pages 204 and 205.

8. a. Which animal might weigh about
   20 times as much as a 30-pound raccoon? _____

   b. Can you name two other animals that might
   weigh 20 times as much as a 30-pound raccoon?

   _____

9. About how many 200-pound American alligators
   weigh about as much as a 3,200-pound
   beluga whale?   _____

Try This

10. Which animal might weigh about 100 times as much as
    the combined weights of a 15-pound
    arctic fox and a 10-pound arctic fox? _____

Math Boxes

---

1. If I wanted to take out a square about 4 times as often as a circle,

I would put in _____ square(s).

2. Put these numbers in order from smallest to largest.

998,752 _____

1,000,008 _____

750,999 _____

1,709,832 _____

SRB 20

---

3. Write equivalent fractions.

$\frac{1}{2} = \dfrac{\square}{\square}$

$\frac{1}{4} = \dfrac{\square}{\square}$

SRB 27–30

4. Pencils cost $1.99 for a package of 24. Estimate. About how much do 4 packages cost?

_____

SRB 191

---

5. Use bills and coins.

Share $45.90 equally among 3 people.

Each gets $_____.

Share $49.20 equally among 4 people.

Each gets $_____.

SRB 73

6. Measure the line segment to the nearest $\frac{1}{2}$ inch.

_____ about _____ in.

Draw a line segment that is $1\frac{1}{2}$ inches long.

SRB 143 144

LESSON 9·2   Mental Multiplication

Solve these problems in your head. Use a slate and chalk, or pencil and paper, to help you keep track of your thinking. For some of the problems, you will need to use the information on journal pages 204 and 205.

1. Could 5 arctic foxes weigh 100 pounds? _____

    Less than 100 pounds? _____

    Explain the strategy you used.

    _____

    _____

    _____

2. Could 12 harp seals weigh more than 1 ton? _____ Less than 1 ton? _____

    Explain the strategy that you used.

    _____

    _____

3. How much do eight 53-pound white-tailed deer weigh? _____

    Explain the strategy that you used.

    _____

    _____

4. How much do six 87-pound sea otters weigh? _____

5. How much do seven 260-pound Atlantic
    green turtles weigh?   _____

6. $6 \times 54 =$ _____   7. _____ $= 4 \times 250$

8. $2 \times 460 =$ _____   9. _____ $= 3 \times 320$

LESSON 9·2 Number Stories

Use the Adult Weights of North American Animals poster on *Math Journal 2,* pages 204 and 205. Make up multiplication and division number stories. Ask a partner to solve your number stories.

1. _____

   _____

   _____

   _____

   _____

   Answer: _____

2. _____

   _____

   _____

   _____

   _____

   Answer: _____

3. _____

   _____

   _____

   _____

   Answer: _____

LESSON 9·2 Math Boxes

1. Nicky has $806 in the bank. Andrew has $589. How much more money does Nicky have than Andrew?

$ _____

SRB 258

2. Write the numbers.
5 tens 9 ones

50 + _____ Total: 59

3 tens 8 ones

_____ + _____ Total: _____

3. _____ hour = 30 minutes

_____ hours = 90 minutes

2 hours = _____ minutes

$1\frac{1}{4}$ hours = _____ minutes

_____ hours = 180 minutes

SRB 247

4. Draw a shape with a perimeter of 14 units.

What is the area of the shape?

_____ square units

SRB 150 151 154 155

5. Fill in the missing numbers. Use fractions.

0          $\frac{2}{3}$

_____   _____ _____

SRB 10 26

6. Circle the most appropriate unit.

length of a calculator:

inches   feet   miles

weight of an adult:

ounces   pounds   tons

amount of gas in a car:

cups   pints   gallons

Array Multiplication 1

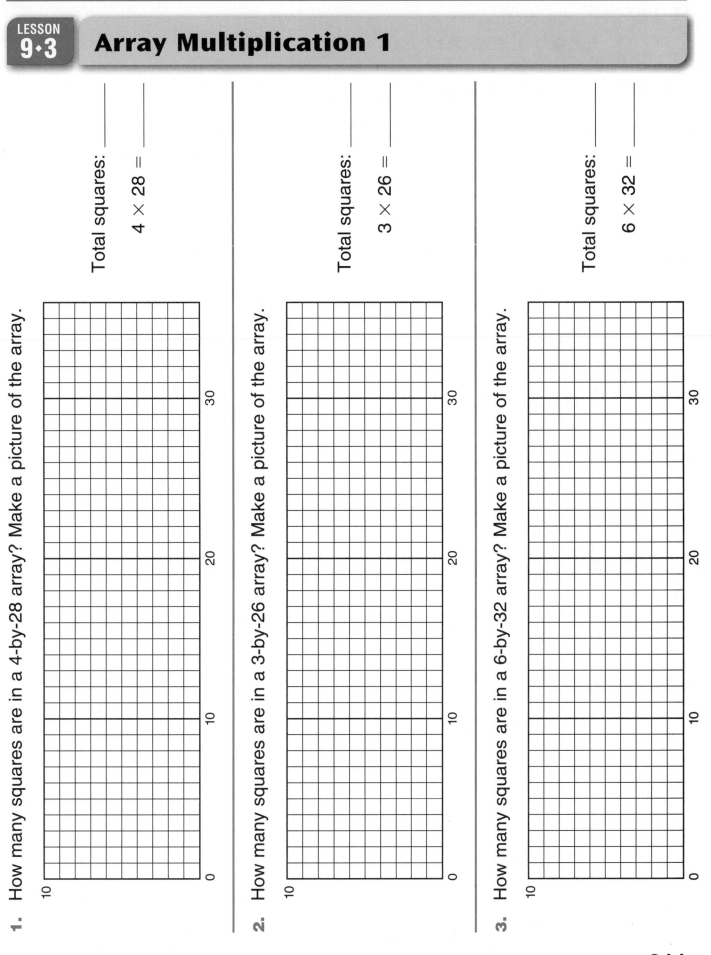

1. How many squares are in a 4-by-28 array? Make a picture of the array.

Total squares: _____

4 × 28 = _____

2. How many squares are in a 3-by-26 array? Make a picture of the array.

Total squares: _____

3 × 26 = _____

3. How many squares are in a 6-by-32 array? Make a picture of the array.

Total squares: _____

6 × 32 = _____

LESSON 9·3 **Geoboard Areas**

Record your results in this table.

Geoboard Areas		
Area	Longer Sides	Shorter Sides
12 square units	units	units
12 square units	units	units
6 square units	units	units
6 square units	units	unit
16 square units	units	units
16 square units	units	units

1. Study your table. Can you find a pattern? _____

_____

_____

2. Find the lengths of the sides of a rectangle or square whose area is 30 square units without using the geoboard or geoboard dot paper. Make or draw the shape to check your answer. _____

3. Make check marks in your table next to the rectangles and squares whose perimeters are 14 units and 16 units.

LESSON 9·3 Math Boxes

1. If I wanted to have an equal chance of taking out a circle or a square, I would add _____ circle(s) to the bag.

2. Which number is the smallest? Fill in the circle for the best answer.

- Ⓐ 1,060
- Ⓑ 1,600
- Ⓒ 1,006
- Ⓓ 6,001

SRB 20

3. Write 3 fractions that are equivalent to $\frac{8}{12}$.

_____ _____ _____

SRB 27–30

4. Pencils cost $1.99 for a package of 24 and $1.69 for a package of 16. What is the total cost of two 24-pencil packages and one 16-pencil package?

Ballpark estimate: _____

Exact answer: _____

SRB 191

5. Use bills and coins.

Share $108 equally among 4 people.

Each gets $_____.

Share $61 equally among 4 people.

Each gets $_____.

SRB 73

6. Measure the line segment to the nearest $\frac{1}{4}$ inch.

_____

about _____ in.

Draw a line segment that is $2\frac{1}{4}$ inches long.

SRB 143 144

LESSON 9·4

Using the Partial-Products Algorithm

Multiply. Compare your answers with a partner. If you disagree, discuss your strategies with each other. Then try the problem again.

Example 7×46	**1.** 34×2

Example 7×46

$$
\begin{array}{r}
46 \\
\times\ 7 \\
\hline
\end{array}
$$

$7\ [40s] \rightarrow\ \ 280$
$7\ [6s] \rightarrow\ +\ 42$
$280 + 42 \rightarrow\ \ 322$

1. 34×2

$$
\begin{array}{r}
34 \\
\times\ 2 \\
\hline
\end{array}
$$

2. 83×5

$$
\begin{array}{r}
83 \\
\times\ 5 \\
\hline
\end{array}
$$

3. 55×6

$$
\begin{array}{r}
55 \\
\times\ 6 \\
\hline
\end{array}
$$

4. 214×7

$$
\begin{array}{r}
214 \\
\times\ \ \ 7 \\
\hline
\end{array}
$$

5. 403×5

$$
\begin{array}{r}
403 \\
\times\ \ \ 5 \\
\hline
\end{array}
$$

LESSON 9·4   **Measures**

Measure these drawings to the nearest $\frac{1}{2}$ inch and $\frac{1}{2}$ centimeter.

1.

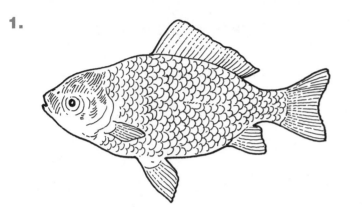

The length of the fish

about _____ in.   about _____ cm

2.   Alpha   Beta

The map distance from Alpha to Beta

about _____ in.   about _____ cm

3.

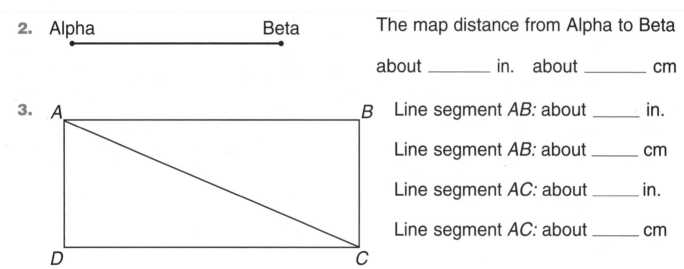

Line segment *AB:* about _____ in.

Line segment *AB:* about _____ cm

Line segment *AC:* about _____ in.

Line segment *AC:* about _____ cm

Try This

Carefully draw the following line segments:

4. 9.5 cm

5. $4\frac{1}{4}$ in.

6. 2 cm shorter than 9.5 cm

7. $1\frac{1}{4}$ in. shorter than $4\frac{1}{4}$ in.

LESSON 9·4 — Math Boxes

1. Morgan earned $252 shoveling snow. Casey earned $228. How much more money did Morgan earn? Fill in the circle for the best answer.

(A) $24   (B) $26

(C) $470   (D) $480

SRB 258

2. Write the numbers.

5 hundreds 6 tens 4 ones

_____ + _____ + _____

Total: _____

3 hundreds 2 tens 9 ones

_____ + _____ + _____

Total: _____

3. _____ seconds = 2 minutes

28 days = _____ weeks

6 months = _____ year

_____ months = $1\frac{1}{2}$ years

SRB 247

4. The length of the longer

side is _____ units.

The length of the shorter

side is _____ units.

The area of the rectangle

is _____ square units.

SRB 154–156

5. Fill in the missing numbers on the number line.

1 $1\frac{3}{4}$

_____   _____ _____

SRB 10 26

6. Circle the unit you would use to measure each item.

weight of journal

ounce   pound   ton

length of football field

inch   yard   mile

length of paperclip

cm   meter   kilometer

LESSON 9·5 Shopping at the Stock-Up Sale

Use the Stock-Up Sale Poster #2 on page 217 in the *Student Reference Book*. Solve each number story below. There is no sales tax. Show how you got the answers.

1. When Mason sees bars of soap at the Stock-Up Sale, he wants to buy at least 5. He has $4.00. Can he buy 5 bars of soap? _____

   Number model: _____

   Can he buy 6 bars? _____

2. Vic's mom gave him a $5.00 bill to buy a toothbrush. If he goes to the sale, can he buy 5 toothbrushes? _____

   Exactly how much money does Vic need in order to be able to buy 5 toothbrushes at the sale price? _____

   Number model: _____

3. Andrea wants 2 bottles of glue. How much more will it cost her to buy 5 bottles at the sale price rather than 2 bottles at the regular price? _____

4. Make up a Stock-Up Sale story of your own.

   _____

   _____

   _____

   Answer: _____

   Number model: _____

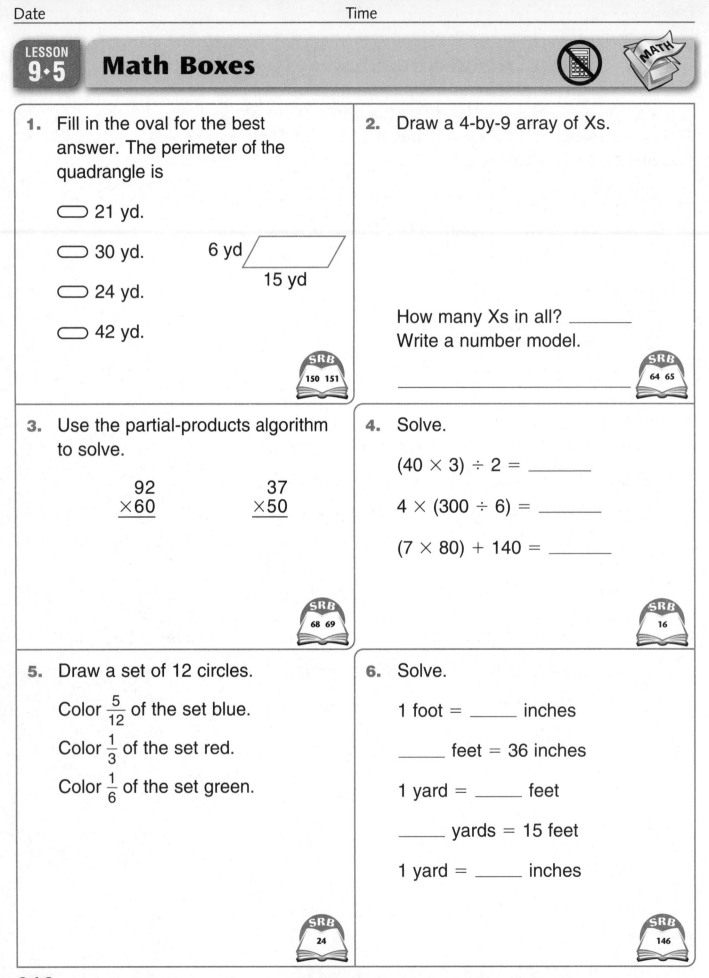

LESSON 9·5   Math Boxes

1. Fill in the oval for the best answer. The perimeter of the quadrangle is

  ◯ 21 yd.

  ◯ 30 yd.   6 yd

  ◯ 24 yd.   15 yd

  ◯ 42 yd.

SRB 150 151

2. Draw a 4-by-9 array of Xs.

How many Xs in all? _____
Write a number model.

_____

SRB 64 65

3. Use the partial-products algorithm to solve.

$$\begin{array}{r} 92 \\ \times 60 \\ \hline \end{array} \qquad \begin{array}{r} 37 \\ \times 50 \\ \hline \end{array}$$

SRB 68 69

4. Solve.

$(40 \times 3) \div 2 =$ _____

$4 \times (300 \div 6) =$ _____

$(7 \times 80) + 140 =$ _____

SRB 16

5. Draw a set of 12 circles.

Color $\frac{5}{12}$ of the set blue.

Color $\frac{1}{3}$ of the set red.

Color $\frac{1}{6}$ of the set green.

SRB 24

6. Solve.

1 foot = _____ inches

_____ feet = 36 inches

1 yard = _____ feet

_____ yards = 15 feet

1 yard = _____ inches

SRB 146

Factor Bingo Game Mat

<table>
<tr><td></td><td></td><td></td><td></td><td></td></tr>
<tr><td></td><td></td><td></td><td></td><td></td></tr>
<tr><td></td><td></td><td></td><td></td><td></td></tr>
<tr><td></td><td></td><td></td><td></td><td></td></tr>
<tr><td></td><td></td><td></td><td></td><td></td></tr>
</table>

Write any of the numbers 2 through 90 onto the grid above.

You may use a number only once.

To help you keep track of the numbers you use, circle them in the list.

	2	3	4	5	6	7	8	9	10
11	12	13	14	15	16	17	18	19	20
21	22	23	24	25	26	27	28	29	30
31	32	33	34	35	36	37	38	39	40
41	42	43	44	45	46	47	48	49	50
51	52	53	54	55	56	57	58	59	60
61	62	63	64	65	66	67	68	69	70
71	72	73	74	75	76	77	78	79	80
81	82	83	84	85	86	87	88	89	90

LESSON 9·6 **Using the Partial-Products Algorithm**

Multiply. Show your work. Compare your answers with your partner's answers. If you disagree, discuss your strategies with each other. Then, try the problem again.

1.
$$\begin{array}{r} 68 \\ \times\ 2 \\ \hline \end{array}$$

2.
$$\begin{array}{r} 96 \\ \times\ 5 \\ \hline \end{array}$$

3.
$$\begin{array}{r} 47 \\ \times\ 4 \\ \hline \end{array}$$

4.
$$\begin{array}{r} 85 \\ \times\ 9 \\ \hline \end{array}$$

5.
$$\begin{array}{r} 231 \\ \times\ 6 \\ \hline \end{array}$$

6.
$$\begin{array}{r} 508 \\ \times\ 5 \\ \hline \end{array}$$

LESSON 9·6 Math Boxes

1. Estimate. Malachi sold 19 boxes of candy for $2.50 a box. About how much money should he have?

about _____

Number model:

_____

SRB 191

2. Solve.

$(9 \times 9) - (43 + 9) =$ _____

_____ $= (5,600 \div 80) \div 2$

_____ $= 963 + (567 - 439)$

SRB 16

3. Use your Fraction Cards. Write $>$, $<$, or $=$ to make the number sentence true.

$\frac{1}{3}$ _____ $\frac{1}{4}$

$\frac{1}{3}$ _____ $\frac{4}{12}$

$\frac{1}{3}$ _____ $\frac{7}{8}$

$\frac{1}{3}$ _____ $\frac{4}{6}$

SRB 31 32

4. Use bills and coins.

Share $63.75 equally among 3 people.

Each gets $_____.

Share $63.00 equally among 5 people.

Each gets $_____.

SRB 73

5. You and a friend are playing a game with a 6-sided die. You win if an odd number is rolled. Your friend wins if an even number is rolled. Do you think this game is fair? Circle one.

yes   no

6. Measure this line segment.

_____

It is about _____ inches long.

It is about _____ centimeters long.

SRB 137–139 143–145

LESSON 9·7   Sharing Money

Work with a partner. Put your play money in a bank for both of you to use.

1. If $54 is shared equally by 3 people, how much does each person get?

   a. How many $10 bills does each person get? _____ $10 bill(s)

   b. How many dollars are left to share? $_____

   c. How many $1 bills does each person get? _____ $1 bill(s)

   d. Number model: $54 ÷ 3 = $_____

2. If $204 is shared equally by 6 people, how much does each person get?

   a. How many $100 bills does each person get? _____ $100 bill(s)

   b. How many $10 bills does each person get? _____ $10 bill(s)

   c. How many dollars are left to share? $_____

   d. How many $1 bills does each person get? _____ $1 bill(s)

   e. Number model: $204 ÷ 6 = $_____

3. If $71 is shared equally by 5 people, how much does each person get?

   a. How many $10 bills does each person get? _____ $10 bill(s)

   b. How many dollars are left to share? $_____

   c. How many $1 bills does each person get? _____ $1 bill(s)

   d. How many $1 bills are left over? _____ $1 bill(s)

   e. If the leftover $1 bill(s) are shared equally,
      how many cents does each person get? $_____

   f. Number model: $71 ÷ 5 = $_____

4. $84 ÷ 3 = $_____

5. $75 ÷ 6 = $_____

6. $181 ÷ 4 = $_____

7. $617 ÷ 5 = $_____

LESSON 9·7 Math Boxes

1. Draw a shape with a perimeter of 20 centimeters.

What is the area of your shape?

_____ square centimeters

SRB 150 151 154 155

2. Draw a 4-by-8 array of Xs.

How many Xs in all? _____
Write a number model.

_____

SRB 64 65

3. Use the partial-products algorithm to solve.

$$\begin{array}{r} 296 \\ \times\ \ 4 \\ \hline \end{array} \qquad \begin{array}{r} 183 \\ \times\ \ 7 \\ \hline \end{array}$$

SRB 68 69

4. Put in the parentheses needed to complete the number sentences.

$15 + 80 \times 90 = 7{,}215$

$14 - 6 \times 800 = 6{,}400$

$60 \times 79 + 1 = 4{,}800$

SRB 16 17

5. What part of this pizza has been eaten?

_____

What part is left?

_____

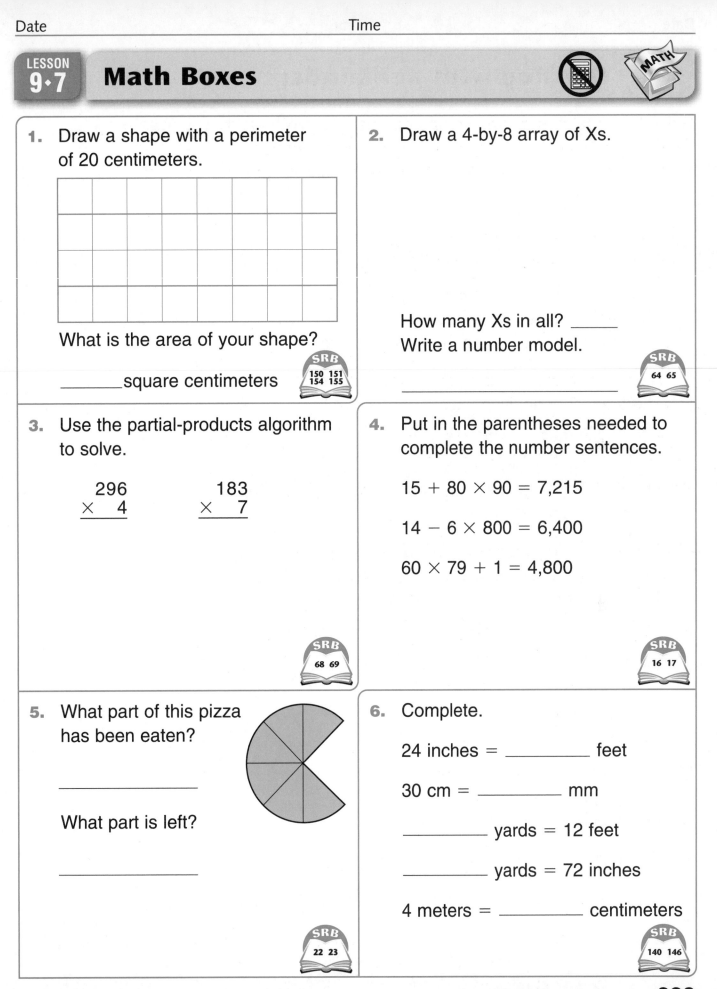

SRB 22 23

6. Complete.

24 inches = _____ feet

30 cm = _____ mm

_____ yards = 12 feet

_____ yards = 72 inches

4 meters = _____ centimeters

SRB 140 146

LESSON 9·8 Division with Remainders

Solve the problems below. Remember that you will have to decide
what the remainder means in order to answer the questions.
You may use your calculator, counters, play money, or pictures.

1. Ruth is buying soda for a party. There are
   6 cans in a pack. She needs 44 cans.
   How many 6-packs will she buy? _____ 6-packs

2. Paul is buying tickets to the circus.
   Each ticket costs $7. He has $47.
   How many tickets can he buy? _____ tickets

3. Héctor is standing in line for the roller coaster.
   There are 33 people in line.
   Each roller coaster car holds 4 people.
   How many cars are needed to hold 33 people? _____ cars

Pretend that the division key on your calculator is broken.
Solve the following problems:

4. Regina is building a fence around her dollhouse.
   She is making each fence post 5 inches tall.
   The wood she bought is 36 inches long.
   How many fence posts does each piece of wood make? _____ posts

   Explain how you found your answer.

   _____

   _____

5. Missy, Ann, and Herman found a $10 bill.
   They want to share the money equally.
   How much money will each person get? _____

   Explain how you found your answer.

   _____

   _____

LESSON 9·8 Math Boxes

1. Kevin and Naomi read to each other for 35 minutes each day. About how many hours do they read to each other in one week?

Answer: about _____
(unit)

2. Fill in the circle for the best answer. $5 \times (6 - 5) =$ _____

Ⓐ 4

Ⓑ 5

Ⓒ 16

Ⓓ 25

SRB
16

3. Write 5 names for $\frac{3}{4}$.

$\frac{3}{4}$

SRB
27–30

4. Use bills and coins.

Share $78 equally among 3 people.

Each person gets $_____._____.

Share $53 equally among 4 people.

Each person gets $_____._____.

SRB
73

5. You and a friend are playing a game with the spinner. You win if the spinner lands on purple. Your friend wins if the spinner lands on black. Do you think this game is fair?

yes

no

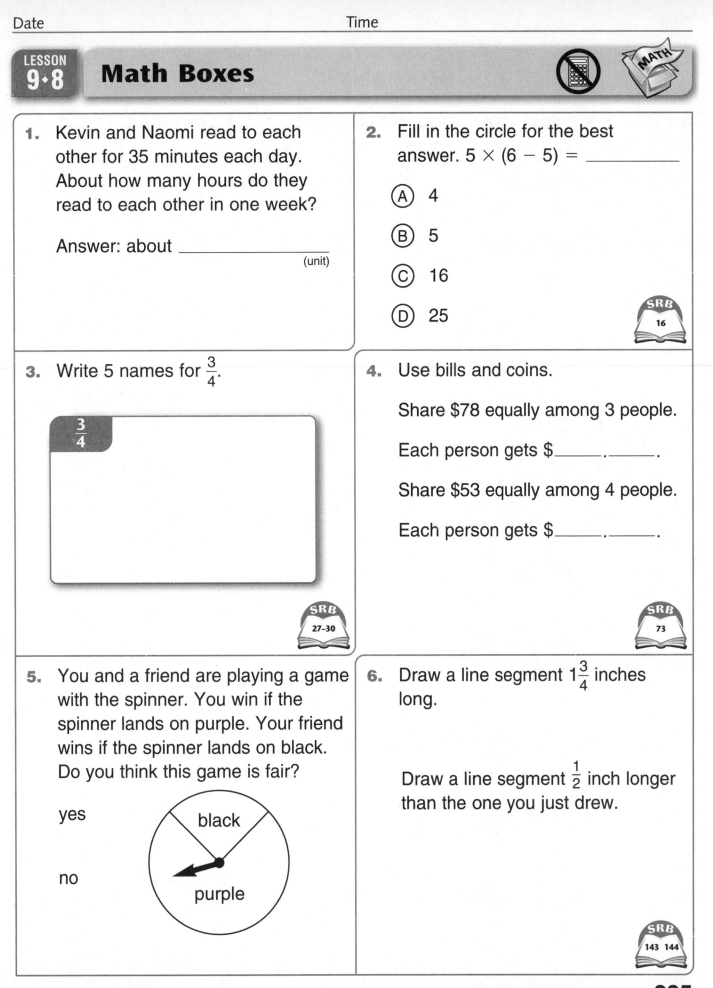

black

purple

6. Draw a line segment $1\frac{3}{4}$ inches long.

Draw a line segment $\frac{1}{2}$ inch longer than the one you just drew.

SRB
143 144

LESSON 9·9 Lattice Multiplication

Megan has a special way of doing multiplication problems. She calls it lattice multiplication. Can you figure out how she does it?

Study the problems and solutions in Column A. Then try to use lattice multiplication to solve the problems in Column B.

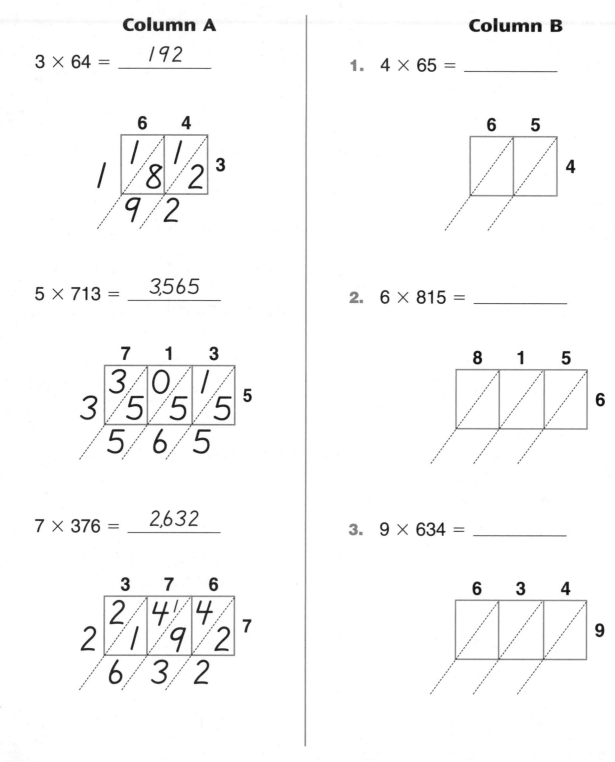

Column A

$3 \times 64 =$ ___192___

$5 \times 713 =$ ___3,565___

$7 \times 376 =$ ___2,632___

Column B

1. $4 \times 65 =$ _____

2. $6 \times 815 =$ _____

3. $9 \times 634 =$ _____

LESSON 9·9 Lattice Multiplication Practice

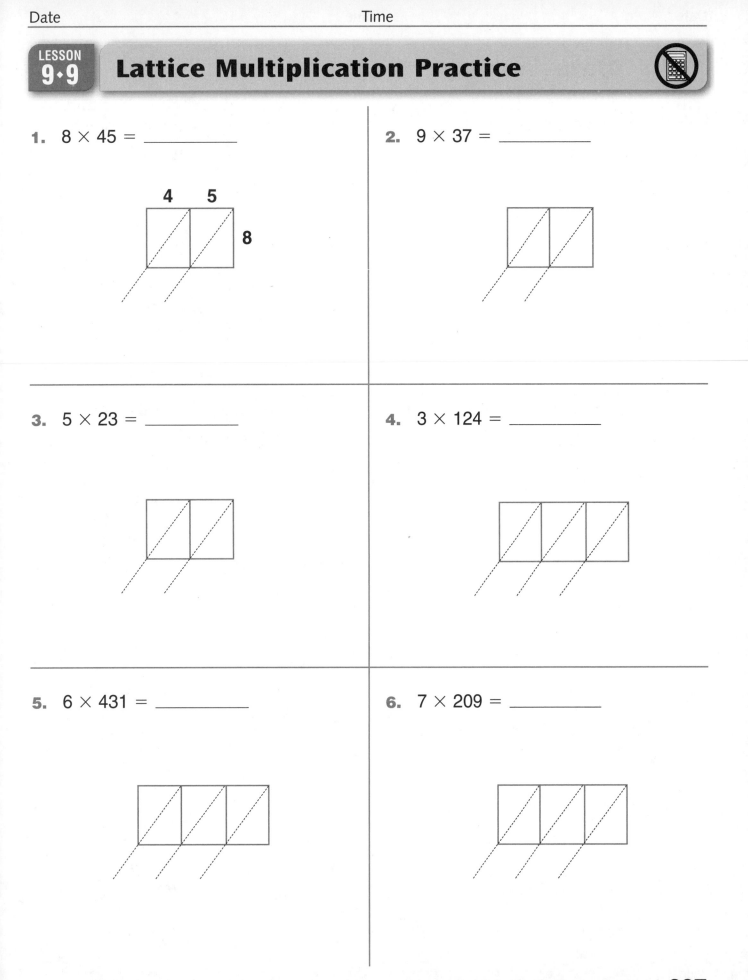

1. 8 × 45 = _____

2. 9 × 37 = _____

3. 5 × 23 = _____

4. 3 × 124 = _____

5. 6 × 431 = _____

6. 7 × 209 = _____

LESSON 9·9 Math Boxes

1. Name the eight factors of 24.

_____, _____, _____, _____,

_____, _____, _____, _____

SRB
37

2. Use the partial-products algorithm to solve. Show your work.

$$238 \times 6 \qquad 574 \times 5$$

SRB
68 69

3. 16 books in all. 3 books per shelf.

How many shelves? _____

How many books
left over?   _____

SRB
74
259 260

4. Draw an angle that measures between 0° and 90°.

SRB
167 168

5. This shape is a _____.

It has _____ sides and _____ vertices.

SRB
103

6. What is the median number of hours children sleep each night?

_____ hours

Hours	Number of Children
8	////
9	⊬⊬ ////
10	////
11	/

SRB
80

LESSON 9·10 Array Multiplication 2

1. How many squares are in a 20-by-13 array?

Total squares = _____

$20 \times 13 =$ _____

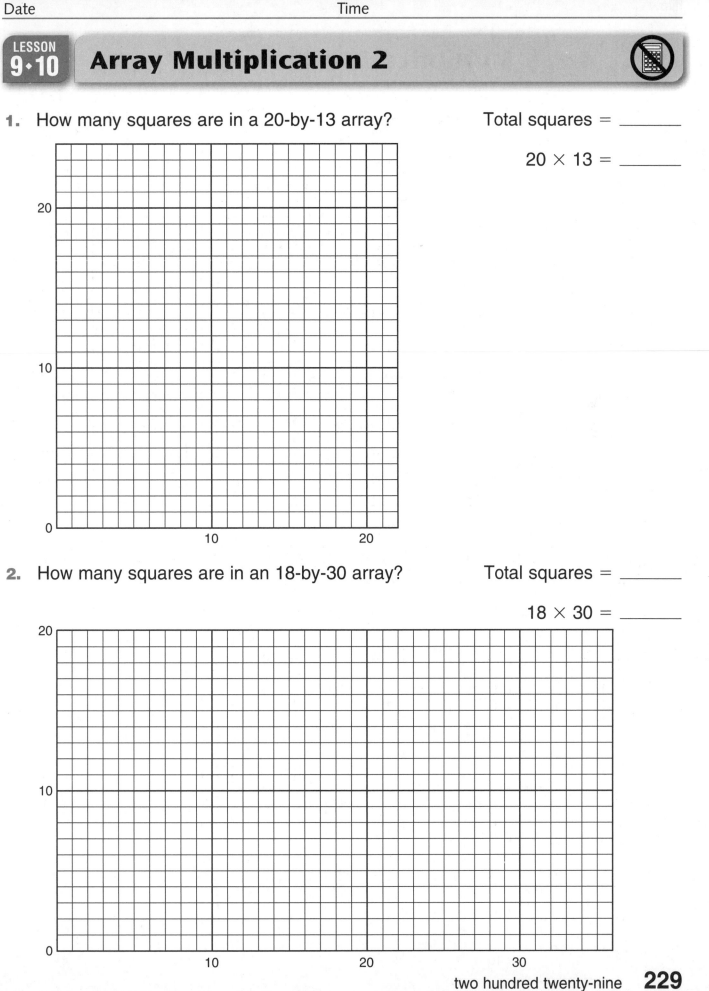

2. How many squares are in an 18-by-30 array?

Total squares = _____

$18 \times 30 =$ _____

LESSON 9·10 Array Multiplication 3

1. How many squares are in a 17-by-34 array?

   Total squares = _____

   $17 \times 34 =$ _____

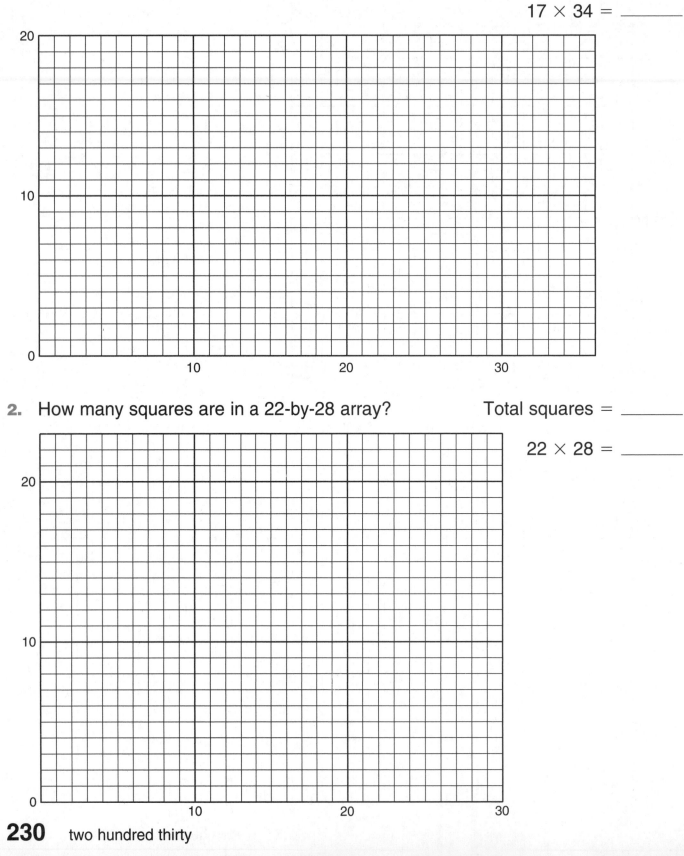

2. How many squares are in a 22-by-28 array?

   Total squares = _____

   $22 \times 28 =$ _____

230 two hundred thirty

LESSON 9·10   Sharing Money

Work with a partner.

Materials   ☐ number cards 0–9 (at least 2 of each)

            ☐ 1 die

            ☐ $10 bills, $1 bills, and tool-kit coins (optional)

Draw 2 number cards. Form a 2-digit number to show how much money will be shared. Roll 1 die to show how many friends will share the money. Fill in the boxes below.

1. $_____ is shared equally by _____ friends.

   a. How many $10 bills does each friend get? _____

   b. How many $1 bills does each friend get? _____

   c. How many $1 bills are left over? _____

   d. If the leftover money is shared equally,
      how many cents does each friend get? _____

   e. Each friend gets a total of $_____.

   f. Number model: _____

Repeat. Draw the next 2 cards. Roll the die. Fill in the blanks below.

2. $_____ is shared equally by _____ friends.

   a. How many $10 bills does each friend get? _____

   b. How many $1 bills does each friend get? _____

   c. How many $1 bills are left over? _____

   d. If the leftover money is shared equally,
      how many cents does each friend get? _____

   e. Each friend gets a total of $_____.

   f. Number model: _____

LESSON 9·10 **Math Boxes**

1. Draw a shape with an area of 10 square centimeters.

What is the perimeter of your shape? _____ centimeters

SRB 150 151 154 155

2. Practice lattice multiplication.

$39 \times 48 =$ _____

SRB 70–72

3. Make an estimate. About how much money, without tax, will you need for 5 gallons of milk that cost $3.09 each?

about _____

SRB 191

4. Estimate. Reuben buys 3 bags of baby carrots at $2.19 per bag. He gives the cashier a $10 bill. About how much change should he get?

about _____

SRB 191

5. Fill in the circle for the best answer. The turn of the angle is

○ **A.** greater than a $\frac{1}{4}$ turn.

○ **B.** less than a $\frac{1}{4}$ turn.

○ **C.** greater than a $\frac{1}{2}$ turn.

○ **D.** a full turn.

SRB 167 168

6. Circle the tool you would use to find the length of a pen:

ruler   compass   scale

the weight of a dime:

ruler   compass   scale

the way to get home:

ruler   compass   scale

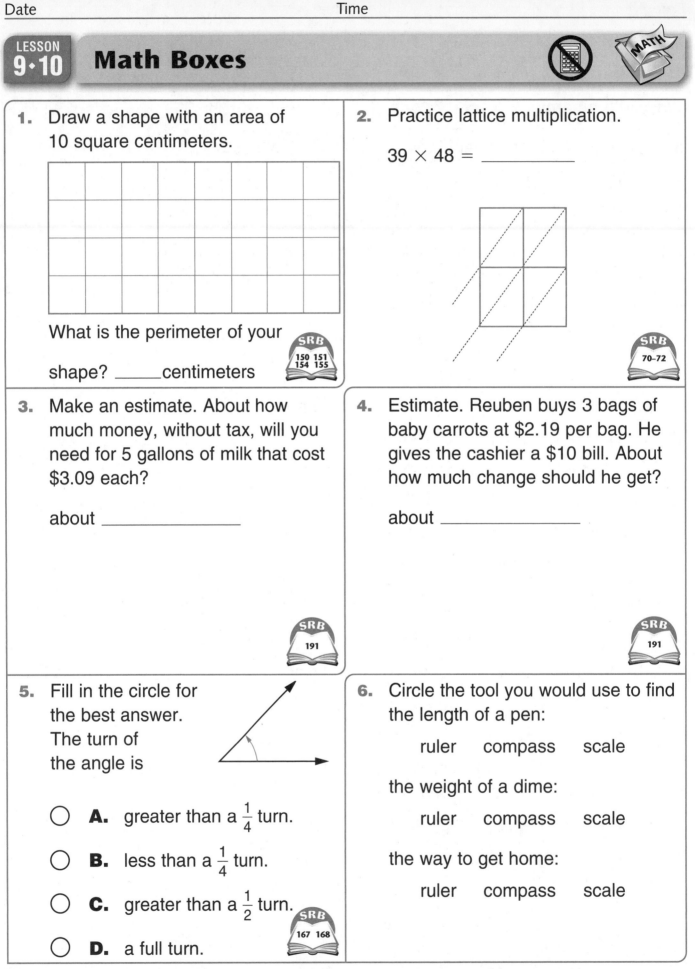

LESSON 9·11 Multiplication with Multiples of 10

Multiply. Compare your answers with a partner. If you disagree,
discuss your strategies with each other. Then try the problem again.

Example: $$\begin{array}{r} 30 \\ \times\ 26 \end{array}$$ 20 [30s]→ 600 6 [30s]→ +180 ────── 780	**1.** $$\begin{array}{r} 70 \\ \times\ 18 \end{array}$$
2. $$\begin{array}{r} 88 \\ \times\ 40 \end{array}$$	**3.** $$\begin{array}{r} 60 \\ \times\ 35 \end{array}$$
4. $$\begin{array}{r} 80 \\ \times\ 44 \end{array}$$	**5.** $$\begin{array}{r} 90 \\ \times\ 63 \end{array}$$

Math Boxes

---

1. Write the six factors of 20.

_____ , _____ , _____ ,

_____ , _____ , _____

SRB
37

---

2. Use the partial-products algorithm
to solve.

$$\begin{array}{r} 489 \\ \times\ \ \ 7 \end{array} \qquad \begin{array}{r} 608 \\ \times\ \ \ 9 \end{array}$$

SRB
68 69

---

3. Allison has 58 stickers. She wants
to share them among 8 friends.

How many stickers
does each friend get?

_____

How many stickers
are left over?

_____

SRB
*73
259 260*

---

4. Fill in the oval for the
best answer. The
degree measure
of the angle is

⬭ less than 40°.

⬭ more than 100°.

⬭ more than 180°.

⬭ 90°.

SRB
167 168

---

5. Fill in the oval for the best answer.
This picture of a 3-dimensional
shape is called a

⬭ rectangular prism.

⬭ pyramid.

⬭ sphere.

It has _____ faces.

SRB
117

---

6. Number of children per classroom:

25, 30, 26, 28, 33, 35, 28

Median: _____

Maximum: _____

Minimum: _____

Range: _____

SRB
79 80

LESSON 9·12

2-Digit Multiplication

Multiply. Compare your answers with a partner. If you disagree, discuss your strategies with each other. Try the problem again.

1.
$$\begin{array}{r} 24 \\ \times\ 16 \\ \hline \end{array}$$

2.
$$\begin{array}{r} 42 \\ \times\ 31 \\ \hline \end{array}$$

3.
$$\begin{array}{r} 12 \\ \times\ 87 \\ \hline \end{array}$$

4.
$$\begin{array}{r} 59 \\ \times\ 79 \\ \hline \end{array}$$

5.
$$\begin{array}{r} 36 \\ \times\ 14 \\ \hline \end{array}$$

6.
$$\begin{array}{r} 42 \\ \times\ 53 \\ \hline \end{array}$$

7. Describe in words how you solved Problem 1.

_____

_____

_____

_____

_____

LESSON 9·12 **Math Boxes**

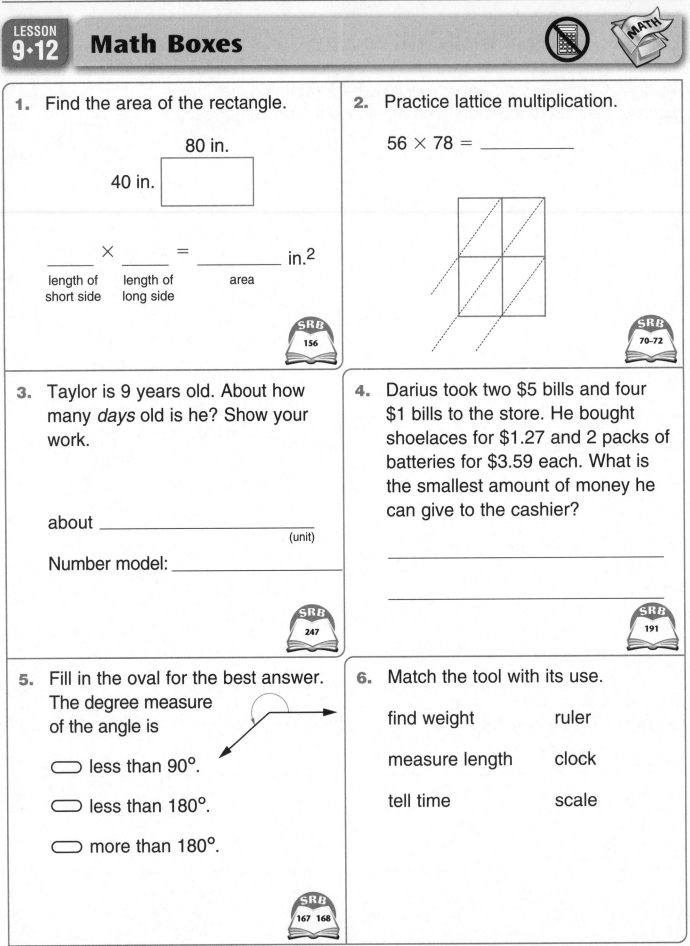

1. Find the area of the rectangle.

   80 in.

   40 in.

   _____ × _____ = _____ in.²
   length of   length of   area
   short side   long side

   SRB 156

2. Practice lattice multiplication.

   56 × 78 = _____

   SRB 70–72

3. Taylor is 9 years old. About how many *days* old is he? Show your work.

   about _____
                          (unit)

   Number model: _____

   SRB 247

4. Darius took two $5 bills and four $1 bills to the store. He bought shoelaces for $1.27 and 2 packs of batteries for $3.59 each. What is the smallest amount of money he can give to the cashier?

   _____

   _____

   SRB 191

5. Fill in the oval for the best answer. The degree measure of the angle is

   ⬭ less than 90°.

   ⬭ less than 180°.

   ⬭ more than 180°.

   SRB 167 168

6. Match the tool with its use.

   find weight   ruler

   measure length   clock

   tell time   scale

LESSON 9·13 — Number Stories with Positive & Negative Numbers

Solve the following problems. Use the thermometer scale, the class number line, or other tools to help.

1. The largest change in temperature in a single day took place in January 1916 in Browning, Montana. The temperature dropped 100°F that day. The temperature was 44°F when it started dropping.

   How low did it go? _____

2. The largest temperature rise in 12 hours took place in Granville, North Dakota, on February 21, 1918. The temperature rose 83°F that day. The high temperature was 50°F.

   What was the low temperature? _____

3. On January 12, 1911, the temperature in Rapid City, South Dakota, fell from 49°F at 6 A.M. to −13°F at 8 A.M.

   By how many degrees did the temperature drop in those 2 hours? _____

4. The highest temperature ever recorded in Verkhoyansk, Siberia, was 98°F. The lowest temperature ever recorded there was −94°F.

   What is the difference between those two temperatures? _____

5. Write your own number story using positive and negative numbers.

_____

_____

_____

°F
100
90
80
70
60
50
40
30
20
10
0
−10
−20
−30
−40
−50
−60
−70
−80
−90
−100

Math Boxes

---

1. length = _____ units

width = _____ units

area = _____ square
units

2 factors of 45
are _____ and _____.

SRB
37
154 156

---

2. Use the partial-products algorithm
to solve.

$$\begin{array}{r} 652 \\ \times 3 \\ \hline \end{array} \qquad \begin{array}{r} 408 \\ \times 8 \\ \hline \end{array}$$

SRB
68 69

---

3. There are 347 candles. A box holds
50 candles. How many full boxes of
candles is that?

_____ boxes

How many candles are left over?

_____ candles

SRB
74
259 260

---

4. Fill in the oval for the best answer.
The degree measure of the angle is

⬭ 180°.

⬭ less than 90°.

⬭ less than 270°.

⬭ more than 270°.

SRB
167 168

---

5. What 3-D shape is this a picture of?
Fill in the oval for the best answer.

⬭ sphere

⬭ cylinder

⬭ pyramid

What is the shape of the base?

_____

SRB
118

---

6. Number of pets children have:

0, 4, 0, 1, 1, 3, 6, 2, 5

Median: _____

Maximum: _____

Minimum: _____

Range: _____

SRB
79 80

---

238 two hundred thirty-eight

LESSON 9·14 Math Boxes

1. Measure this line segment to the nearest $\frac{1}{8}$ inch.

_____

It is about _____ inches long.

Draw a line segment $1\frac{3}{4}$ inches long.

SRB 143–145

2. Measure this line segment to the nearest $\frac{1}{2}$ centimeter.

_____

It is about _____ centimeters long.

Draw a line segment 3.5 centimeters long.

SRB 137–139

3. Circle the most appropriate unit.

length of calculator

inches   feet   miles

weight of a third grader

ounces   pounds   tons

amount of water in a drinking glass

cups   quarts   gallons

4. Find the median of the following numbers.

34, 56, 34, 16, 33, 27, 45

Median: _____

SRB 80

5. Find the maximum, minimum, and range of the following numbers:

18, 13, 6, 9, 15, 25, 21, 17

Maximum: _____

Minimum: _____

Range: _____

SRB 79

6. Circle the tool you use to find

the temperature:

scale   thermometer   ruler

the weight of a *Student Reference Book*:

scale   thermometer   ruler

the perimeter of a *Student Reference Book*:

scale   thermometer   ruler

LESSON 10·1 **Review: Units of Measure**

1. Measure in centimeters. Which is longer, the path from *A* to *B* or the path from *C* to *D*? _____

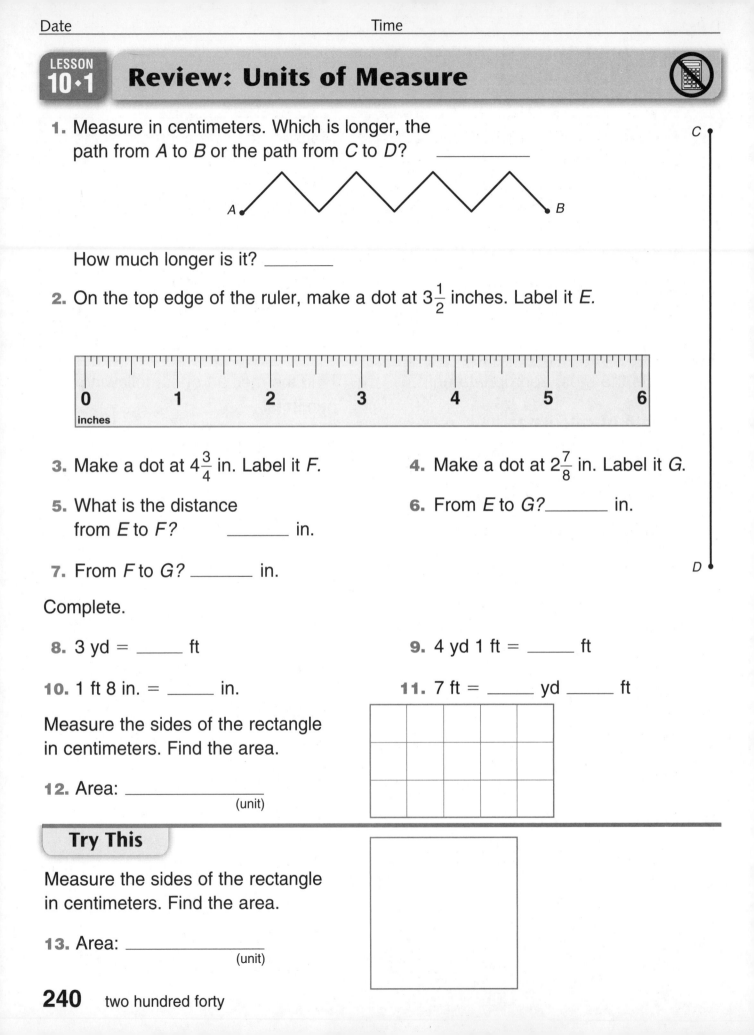

How much longer is it? _____

2. On the top edge of the ruler, make a dot at $3\frac{1}{2}$ inches. Label it *E*.

3. Make a dot at $4\frac{3}{4}$ in. Label it *F*.

4. Make a dot at $2\frac{7}{8}$ in. Label it *G*.

5. What is the distance from *E* to *F*? _____ in.

6. From *E* to *G*? _____ in.

7. From *F* to *G*? _____ in.

Complete.

8. 3 yd = _____ ft

9. 4 yd 1 ft = _____ ft

10. 1 ft 8 in. = _____ in.

11. 7 ft = _____ yd _____ ft

Measure the sides of the rectangle in centimeters. Find the area.

12. Area: _____
(unit)

Try This

Measure the sides of the rectangle in centimeters. Find the area.

13. Area: _____
(unit)

LESSON 10·1

Weight and Volume

Complete Part 1 before the start of Lesson 10-3.

Part 1 Order the objects on display from heaviest (1) to lightest (4). Lift them to help you guess. Record your guesses below.

1. _____   2. _____

3. _____   4. _____

Complete Part 2 as part of Lesson 10-3.

Part 2 Record the actual order of the objects from heaviest (1) to lightest (4). Were your guesses correct?

1. _____   2. _____

3. _____   4. _____

Complete Parts 3 and 4 as part of Lesson 10-4.

Part 3 Order the objects on display from largest (1) to smallest (4) volume. Record your guesses below.

1. _____   2. _____

3. _____   4. _____

Part 4 Record the actual order of the objects from largest (1) to smallest (4) volume. Were your guesses correct?

1. _____   2. _____

3. _____   4. _____

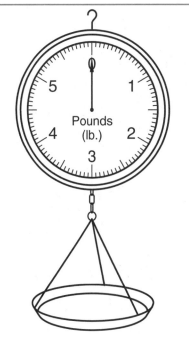

1. Make a dot on the produce scale to show 2 lb. Label it *A.*

2. Make a dot on the produce scale to show $3\frac{1}{2}$ lb. Label it *B.*

3. Make a dot on the produce scale to show 2 lb, 8 oz. Label it *C.*

SRB 68–72

LESSON 10·1 Multiplication Practice

Use your favorite multiplication algorithm to solve the following problems.
Show your work.

1.
$$\begin{array}{r} 427 \\ \times \quad 3 \\ \hline \end{array}$$

2.
$$\begin{array}{r} 505 \\ \times \quad 8 \\ \hline \end{array}$$

3.
$$\begin{array}{r} 20 \\ \times \ 90 \\ \hline \end{array}$$

4.
$$\begin{array}{r} 67 \\ \times \ 40 \\ \hline \end{array}$$

5.
$$\begin{array}{r} 74 \\ \times \ 35 \\ \hline \end{array}$$

6.
$$\begin{array}{r} 37 \\ \times \ 58 \\ \hline \end{array}$$

LESSON 10·1 Math Boxes

1. Measure the line segment to the nearest $\frac{1}{2}$ centimeter.

   _____

   _____

   SRB
   137–139

2. Circle the units you would use to measure each item.

   height of a third grader
       inches   miles   square yards

   length of a basketball court
       miles   feet   inches

3. Fill in the circle next to the picture of the triangular prism.

   A ◯

   B ◯

   C ◯

   D ◯

   SRB
   115

4. Rectangle *HFCD* is a(n)

   ____-by-____ rectangle.

   The area of rectangle *HFCD:*

   ____ × ____ = _____ square units.

   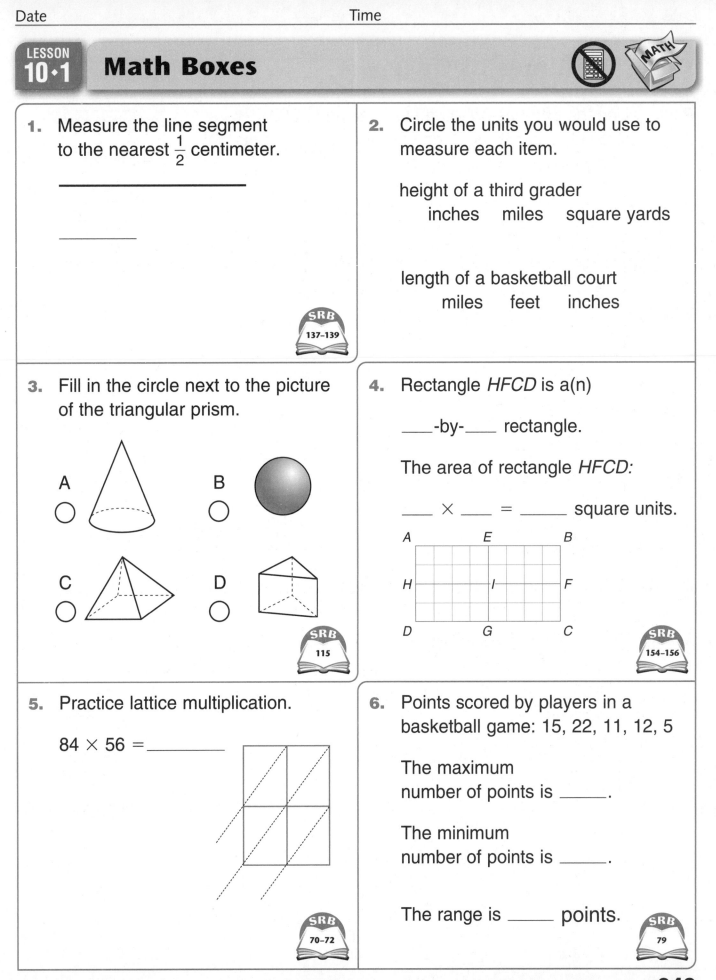

   A   E   B
   H   I   F
   D   G   C

   SRB
   154–156

5. Practice lattice multiplication.

   $84 \times 56 =$ _____

   SRB
   70–72

6. Points scored by players in a basketball game: 15, 22, 11, 12, 5

   The maximum
   number of points is _____.

   The minimum
   number of points is _____.

   The range is _____ points.

   SRB
   79

LESSON 10·2 Volumes of Boxes

Part 1 Use the patterns on *Math Masters,* page 323 to build Boxes A, B, C, and D. Record the results in the table.

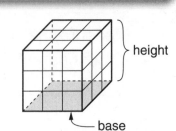

height

base

Box	Number of cm Cubes		Area of Base (square cm)	Height (cm)	Volume (cubic cm)
	Estimate	Exact			
A					
B					
C					
D					

Part 2 The following patterns are for Boxes E, F, and G. Each square stands for 1 square centimeter. Find the volume of each box. (Do not cut out the patterns.)

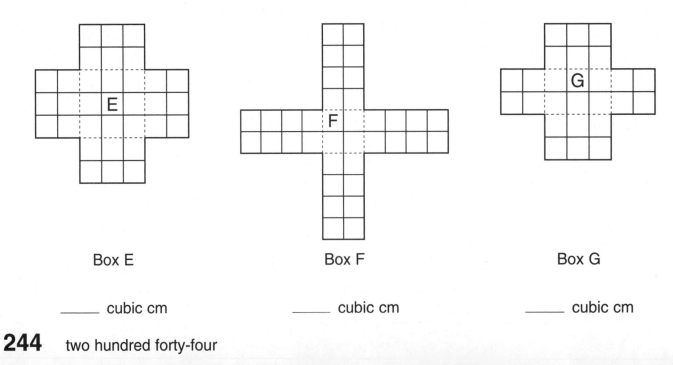

Box E

Box F

Box G

_____ cubic cm   _____ cubic cm   _____ cubic cm

LESSON 10·2 Math Boxes

1. Circle any measurements in Column B that match the one in Column A.

Column A	Column B	
2 feet	12 in.	3 yd
	24 in.	1 yd
3 feet	36 in.	1 m
	1 yd	30 in.
2 yards	50 in.	72 in.
	6 ft	9 ft

SRB 146

2. Use the partial-products algorithm to solve.

$$\begin{array}{r} 86 \\ \times\ 27 \\ \hline \end{array} \qquad \begin{array}{r} 91 \\ \times\ 64 \\ \hline \end{array}$$

SRB 68 69

3. There are 20 crayons in a box. $\frac{1}{2}$ of the crayons are broken. How many crayons are broken?

_____ crayons

$\frac{1}{4}$ of the crayons are red.

How many crayons are red?

_____

SRB 24

4. Fill in the circle next to the numbers that are in order from smallest to largest.

Ⓐ 0, 6, −3, 0.15

Ⓑ 6, 0.15, 0, −3

Ⓒ 0.15, 0, −3, 6

Ⓓ −3, 0, 0.15, 6

SRB 33–35

5. Write the number that has

2 in the thousandths place
6 in the ones place
3 in the hundredths place
4 in the tenths place

___ . ___ ___ ___

SRB 35

6. Complete the "What's My Rule?" table.

in →

Rule
Add 25 minutes

→ out

in	out
7:00	
3:15	
5:45	
	7:40
	11:10

SRB 203 204

LESSON 10·3 **Various Scales**

Refer to pages 165 and 166 in your *Student Reference Book*. For each scale shown, list three things you could weigh on the scale.

balance scale

_____

_____

_____

market scale

_____

_____

_____

package scale

_____

_____

_____

bath scale

_____

_____

_____

spring scale

_____

_____

_____

produce scale

_____

_____

_____

letter scale

_____

_____

_____

platform scale

_____

_____

_____

infant scale

_____

_____

_____

diet/food scale

_____

_____

_____

LESSON 10·3   Reading Scales

Read each scale and record the weight.

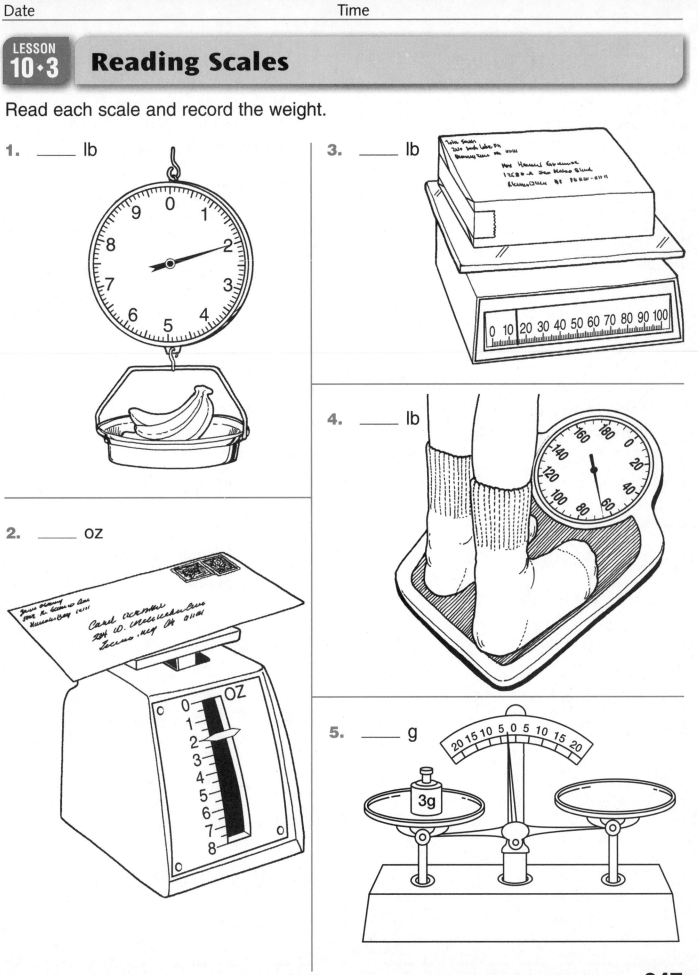

1. _____ lb

2. _____ oz

3. _____ lb

4. _____ lb

5. _____ g

LESSON 10·3 **Math Boxes**

1. Measure the line segment to the nearest $\frac{1}{2}$ inch.

_____

_____

SRB
143–145

2. Circle the units you would use to measure each item.

length of a swimming pool
    meters   kilometers   centimeters

length of an ant
    meters   kilometers   millimeters

3. This is a picture of a 3-dimensional shape. Name the shape.

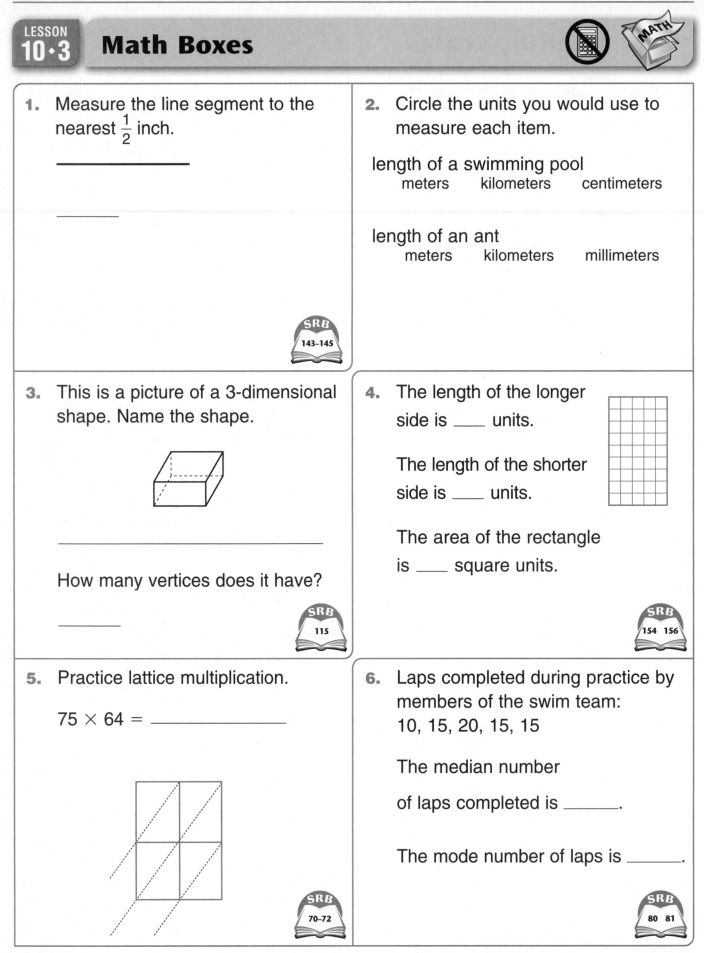

_____

How many vertices does it have?

_____

SRB
115

4. The length of the longer side is ____ units.

The length of the shorter side is ____ units.

The area of the rectangle is ____ square units.

SRB
154 156

5. Practice lattice multiplication.

$75 \times 64 =$ _____

SRB
70–72

6. Laps completed during practice by members of the swim team:
10, 15, 20, 15, 15

The median number of laps completed is _____.

The mode number of laps is _____.

SRB
80 81

Math Boxes

1. Write equivalent lengths.

$\frac{1}{3}$ yd = _____ ft

18 in. = _____ yd

50 mm = _____ cm

0.6 m = _____ cm

SRB
140
146 147

2. Use the partial-products algorithm to solve. Show your work.

$$\begin{array}{r} 36 \\ \times\ 25 \\ \hline \end{array} \qquad \begin{array}{r} 43 \\ \times\ 65 \\ \hline \end{array}$$

SRB
68 69

3. Complete the fraction number story.

Samantha ate $\dfrac{\boxed{}}{8}$ of the pizza.

Luke ate $\dfrac{\boxed{}}{8}$ of the pizza.

Connor ate $\dfrac{\boxed{}}{8}$ of the pizza.

$\dfrac{\boxed{}}{8}$ of the pizza was left over.

SRB
22 23

4. Find the distance between each pair of numbers.

2 and −6 _____

−7 and 15 _____

100 and −500 _____

SRB
39 40

5. In the number 42.368:

the 3 means ___ *3 tenths* ___

the 2 means _____

the 8 means _____

the 6 means _____

the 4 means _____

SRB
35

6. Tatiana gets her teeth cleaned every 6 months. If her last appointment was in February, when is her next appointment?

_____

SRB
176 177

LESSON 10·5 Units of Measure

Mark the unit you would use to measure each item.

1. thickness of a dime   ⬭ millimeter   ⬭ gram   ⬭ foot

2. flour used in cooking   ⬭ gallon   ⬭ cup   ⬭ liter

3. gasoline for a car   ⬭ fluid ounce   ⬭ ton   ⬭ gallon

4. distance to the moon   ⬭ foot   ⬭ square mile   ⬭ kilometer

5. area of a floor   ⬭ square foot   ⬭ cubic foot   ⬭ foot

6. draperies   ⬭ kilometer   ⬭ millimeter   ⬭ yard

7. diameter of a basketball   ⬭ mile   ⬭ inch   ⬭ square inch

8. perimeter of a garden   ⬭ yard   ⬭ square yard   ⬭ centimeter

9. spices in a recipe   ⬭ teaspoon   ⬭ pound   ⬭ fluid ounce

10. weight of a nickel   ⬭ pound   ⬭ gram   ⬭ inch

11. volume of a suitcase   ⬭ square inch   ⬭ foot   ⬭ cubic inch

12. length of a cat's tail   ⬭ centimeter   ⬭ meter   ⬭ yard

Mark the best answer.

13. How much can an 8-year-old grow in a year?

     ⬭ about 2 in.   ⬭ about 2 ft   ⬭ about 1 yd   ⬭ about 1 m

14. How long would it take you to walk 3 miles?

     ⬭ about 10 min   ⬭ about 20 min   ⬭ about 1 hour   ⬭ about 5 hours

Body Measures

Work with a partner to make each measurement to the nearest $\frac{1}{2}$ inch.

	Adult at Home	Me (Now)	Me (Later)
Date			
height	about _____ in.	about _____ in.	about _____ in.
shoe length	about _____ in.	about _____ in.	about _____ in.
around neck	about _____ in.	about _____ in.	about _____ in.
around wrist	about _____ in.	about _____ in.	about _____ in.
waist to floor	about _____ in.	about _____ in.	about _____ in.
forearm	about _____ in.	about _____ in.	about _____ in.
hand span	about _____ in.	about _____ in.	about _____ in.
arm span	about _____ in.	about _____ in.	about _____ in.
_____	about _____ in.	about _____ in.	about _____ in.
_____	about _____ in.	about _____ in.	about _____ in.

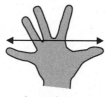

forearm   hand span   arm span

LESSON
10·5
Math Boxes

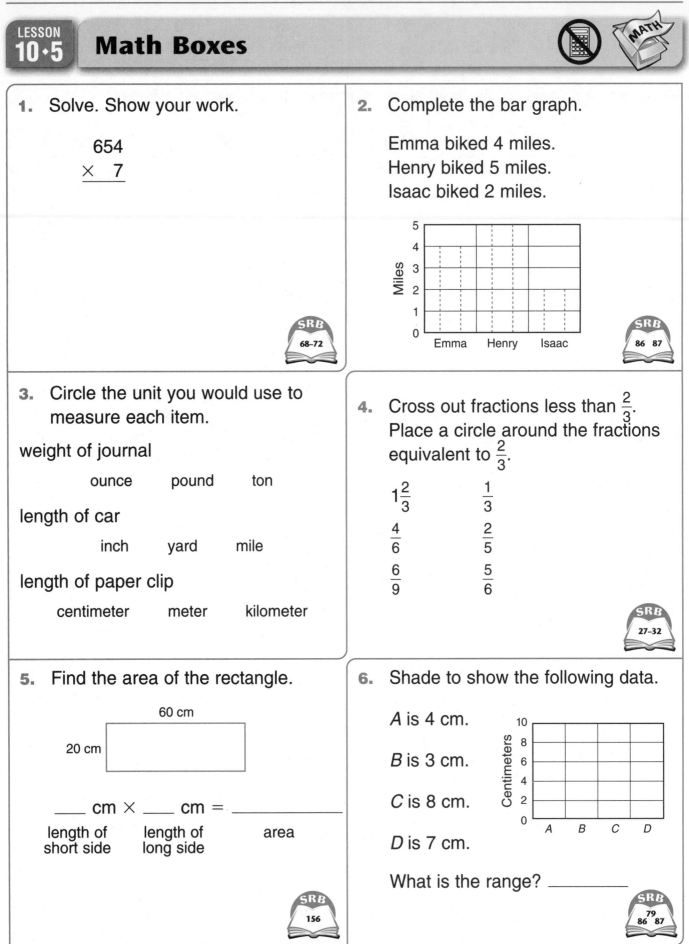

1. Solve. Show your work.

654
×   7

SRB
68–72

2. Complete the bar graph.

Emma biked 4 miles.
Henry biked 5 miles.
Isaac biked 2 miles.

SRB
86 87

3. Circle the unit you would use to measure each item.

weight of journal

ounce   pound   ton

length of car

inch   yard   mile

length of paper clip

centimeter   meter   kilometer

4. Cross out fractions less than $\frac{2}{3}$. Place a circle around the fractions equivalent to $\frac{2}{3}$.

$1\frac{2}{3}$ $\frac{1}{3}$

$\frac{4}{6}$ $\frac{2}{5}$

$\frac{6}{9}$ $\frac{5}{6}$

SRB
27–32

5. Find the area of the rectangle.

60 cm

20 cm

____ cm × ____ cm = _____
length of   length of   area
short side   long side

SRB
156

6. Shade to show the following data.

A is 4 cm.

B is 3 cm.

C is 8 cm.

D is 7 cm.

What is the range? _____

SRB
79
86 87

A Mean, or Average, Number of Children

LESSON 10·6

Activity 1 Make a bar graph of the data in the table.

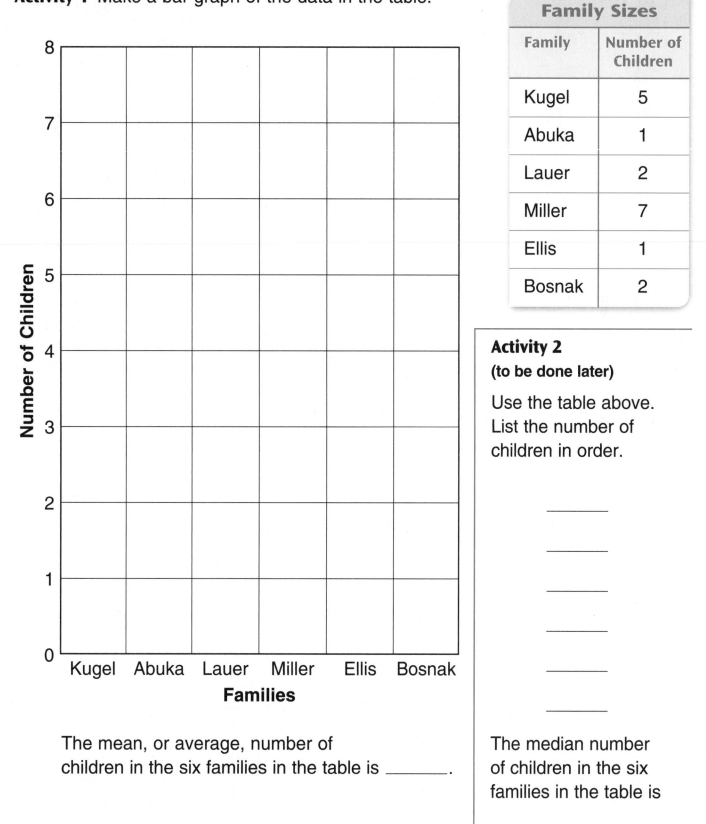

Family Sizes

Family	Number of Children
Kugel	5
Abuka	1
Lauer	2
Miller	7
Ellis	1
Bosnak	2

Y-axis: **Number of Children** (0–8)
X-axis: **Families** (Kugel, Abuka, Lauer, Miller, Ellis, Bosnak)

The mean, or average, number of children in the six families in the table is _____.

Activity 2
(to be done later)

Use the table above. List the number of children in order.

_____

_____

_____

_____

_____

_____

The median number of children in the six families in the table is

_____.

LESSON 10·6 A Mean, or Average, Number of Eggs

Activity 1 Make a bar graph of the data in the table.

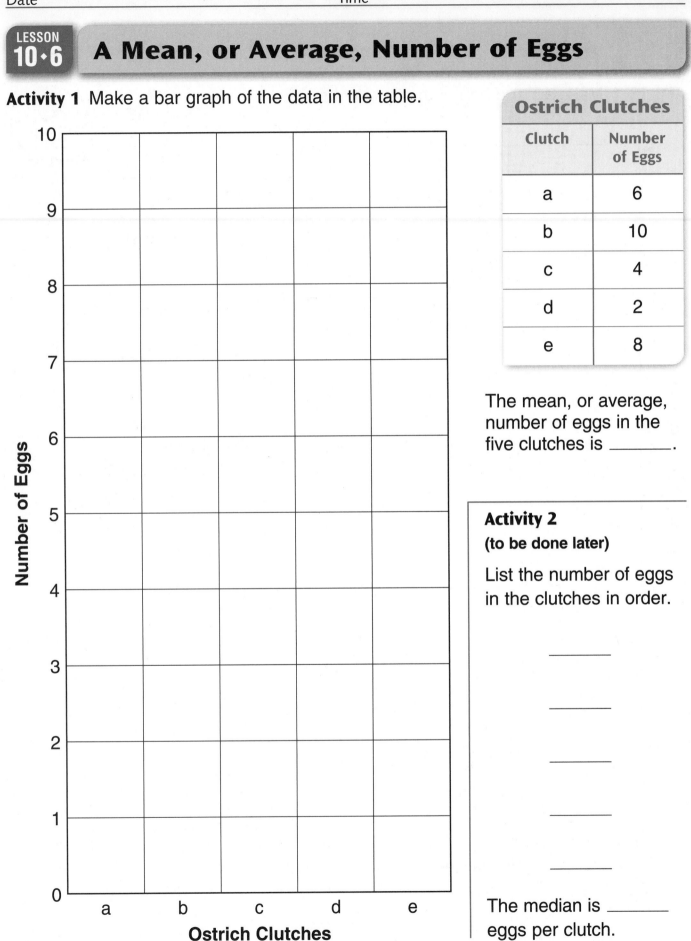

Number of Eggs (y-axis: 0–10)

Ostrich Clutches (x-axis: a b c d e)

Ostrich Clutches

Clutch	Number of Eggs
a	6
b	10
c	4
d	2
e	8

The mean, or average, number of eggs in the five clutches is _____.

Activity 2
(to be done later)

List the number of eggs in the clutches in order.

_____

_____

_____

_____

_____

The median is _____ eggs per clutch.

LESSON 10·6 **Math Boxes**

1. Measure each side of the triangle to the nearest centimeter.

_____ cm

_____ cm

_____ cm

Perimeter = _____ cm

SRB 137–139

2. There are 5 blocks in a bag. 2 blocks are red, 2 blocks are blue, and 1 block is green. What are the chances of pulling out a red block?

_____ out of _____ chances

SRB 94

3. James built a rectangular prism out of base-10 blocks. He used 30 cm cubes to make the base. He put 4 more layers of cubes on top of that. What is the volume of the prism he built?

_____ cubic centimeters

SRB 157–159

4. Complete.

1 gallon = _____ quarts

_____ gallons = 12 quarts

1 pint = _____ cups

_____ pints = 14 cups

1 cup = _____ fl oz

_____ cups = 72 fl oz

SRB 160 161

5. Molly is playing with 5 toy cars. This is only $\frac{1}{3}$ of her set of cars. How many cars are in her complete set? Fill in the circle next to the best answer.

(A) $\frac{5}{3}$ cars   (C) 10 cars

(B) 5 cars   (D) 15 cars

SRB 24

6. Color the circle so that it matches the description.

$\frac{1}{2}$ blue

$\frac{1}{3}$ green

$\frac{1}{6}$ yellow

Which color would you expect the spinner to land on most often? _____

SRB 93

LESSON 10·7   Finding the Median and the Mean

1. The median (middle) arm span in my class is about _____ inches.

2. The mean (average) arm span in my class is about _____ inches.

3. Look at page 251 in your journal. Use the measurements for an adult and the *second* measurements for yourself to find the median and mean arm spans and heights for your group. Record the results in the table below.

   a. Find the median and mean arm spans of the *adults* for your group.

   b. Find the median and mean arm spans of the *children* for your group.

   c. Find the median and mean heights of the *adults* for your group.

   d. Find the median and mean heights of the *children* for your group.

Summary of Measurements for Your Group		
Measure	**Adults**	**Children**
Median arm span		
Mean arm span		
Median height		
Mean height		

Find the mean of each set of data. Use your calculator.

4. High temperatures: 56°F, 62°F, 74°F, 68°F   _____°F

5. Low temperatures: 32°F, 42°F, 58°F, 60°F   _____°F

6. Ticket sales: $710, $650, $905 $_____

7. Throws: 40 ft, 32 ft, 55 ft, 37 ft, 43 ft, 48 ft   _____ ft

LESSON 10·7 **Math Boxes**

---

1. Solve. Show your work.

$$\begin{array}{r} 837 \\ \times \quad 4 \\ \hline \end{array}$$

SRB 68–72

---

2. Read the graph. **Days of Rain**
Which month had the most days of rain?

_____

What is the median number of days of rain? _____

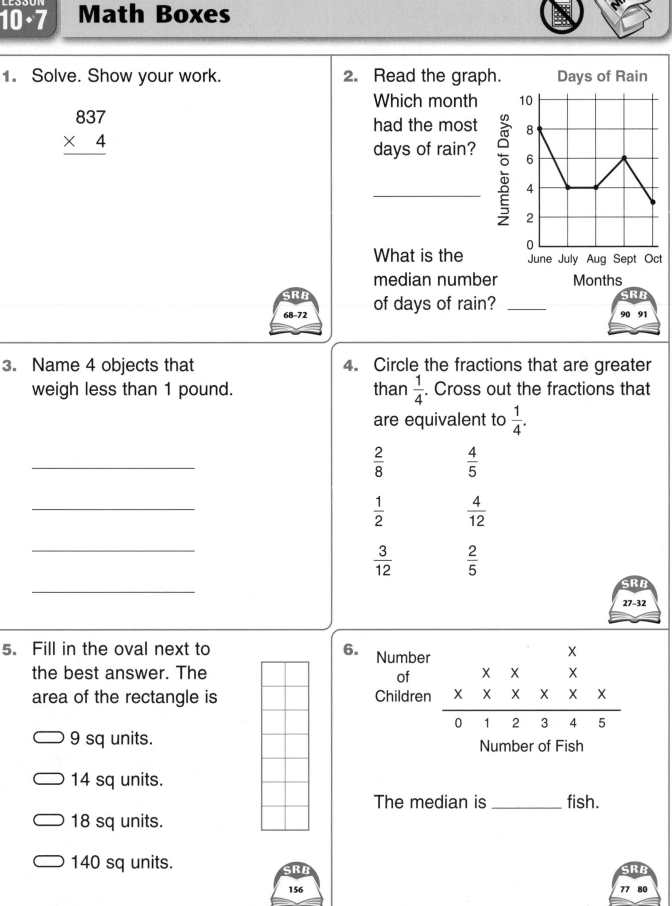

SRB 90 91

---

3. Name 4 objects that weigh less than 1 pound.

_____

_____

_____

_____

---

4. Circle the fractions that are greater than $\frac{1}{4}$. Cross out the fractions that are equivalent to $\frac{1}{4}$.

$\frac{2}{8}$ $\qquad$ $\frac{4}{5}$

$\frac{1}{2}$ $\qquad$ $\frac{4}{12}$

$\frac{3}{12}$ $\qquad$ $\frac{2}{5}$

SRB 27–32

---

5. Fill in the oval next to the best answer. The area of the rectangle is

⬭ 9 sq units.

⬭ 14 sq units.

⬭ 18 sq units.

⬭ 140 sq units.

SRB 156

6.

Number of Children

```
                              X
                    X   X     X
          X   X   X   X   X   X
          0   1   2   3   4   5
              Number of Fish
```

The median is _____ fish.

SRB 77 80

LESSON 10·8   Calculator Memory

For each problem:

◆ Clear the display and memory:
   Press [MRC] [MRC] [ON/C] or [AC].

◆ Enter the problem key sequence.

◆ Guess what number is in memory.

◆ Record your guess.

◆ Check your guess. Press [MRC] or [MR].

◆ Record the answer.

Your display should look like this

[0.]

before you start a new problem.

Problem Key Sequence	Your Guess	Number in Memory
1. 7 [M+] 9 [M+]		
2. 20 [M+] 16 [M-]		
3. 5 [M+] 10 [M+] 20 [M+]		
4. 12 [+] 8 [M+] 6 [M-]		
5. 15 [−] 9 [M+] 2 [M-]		
6. 25 [M+] 7 [+] 8 [M-]		
7. 30 [M+] 15 [−] 5 [M+]		
8. 2 [+] 2 [M+] 2 [−] 2 [M+] 2 [+] 2 [M+]		

9. Make up your own key sequences with memory keys. Ask a partner to enter the key sequence, guess what number is in memory, and then check the answer.

Problem Key Sequence	Your Guess	Number in Memory

LESSON 10·8 Measurement Number Stories

1. The gas tank of Mrs. Rone's car holds about 12 gallons. About how many gallons are in the tank when the gas gauge shows the tank to be $\frac{3}{4}$ full?

workspace

_____

2. When the gas tank of Mrs. Rone's car is about half empty, she stops to fill the tank. If gas costs $1.25 per gallon, about how much does it cost to fill the tank?

_____

Harry's room measures 11 feet by 13 feet. The door to his room is 3 feet wide. He wants to put a wooden border, or baseboard, around the base of the walls.

3. Draw a diagram of Harry's room on the grid below. Show where the door is. Let each side of a grid square equal 1 foot.

4. How many feet of baseboard must Harry buy? _____

5. How many yards is that? _____

6. If baseboard costs $4.00 a yard, how much will Harry pay? _____

workspace

door

LESSON 10·8   Math Boxes

1. Measure each side of the quadrangle to the nearest half-centimeter.

_____ cm

_____ cm   _____ cm

_____ cm

Another name for this quadrangle is a _____.

SRB 108 109 137–139

2. Fill in the oval for the best answer. There are 6 blocks in a bag. 5 blocks are blue and 1 block is red. The chances of drawing the red block are:

◯ 1 out of 6.

◯ 5 out of 6.

◯ 1 out of 5.

◯ 5 out of 11.

SRB 94

3. Chanel built a rectangular prism out of base-10 blocks. She used 50 cm cubes to make the base. She put 9 more layers of cubes on top of that. What is the volume of the prism she built?

_____ cubic centimeters

SRB 157–159

4. 1 quart = _____ pints

_____ quarts = 16 pints

1 quart = _____ fl oz

_____ quarts = 96 fl oz

1 gallon = _____ fl oz

SRB 160 161

5. There are 24 children in Mrs. Hiller's class. $\frac{1}{2}$ of the children play soccer. How many children play soccer?

_____ children

$\frac{1}{3}$ of the children play a musical instrument. How many children play a musical instrument?

_____ children

SRB 24

6. Design a spinner that has an equal chance of landing on red or green.

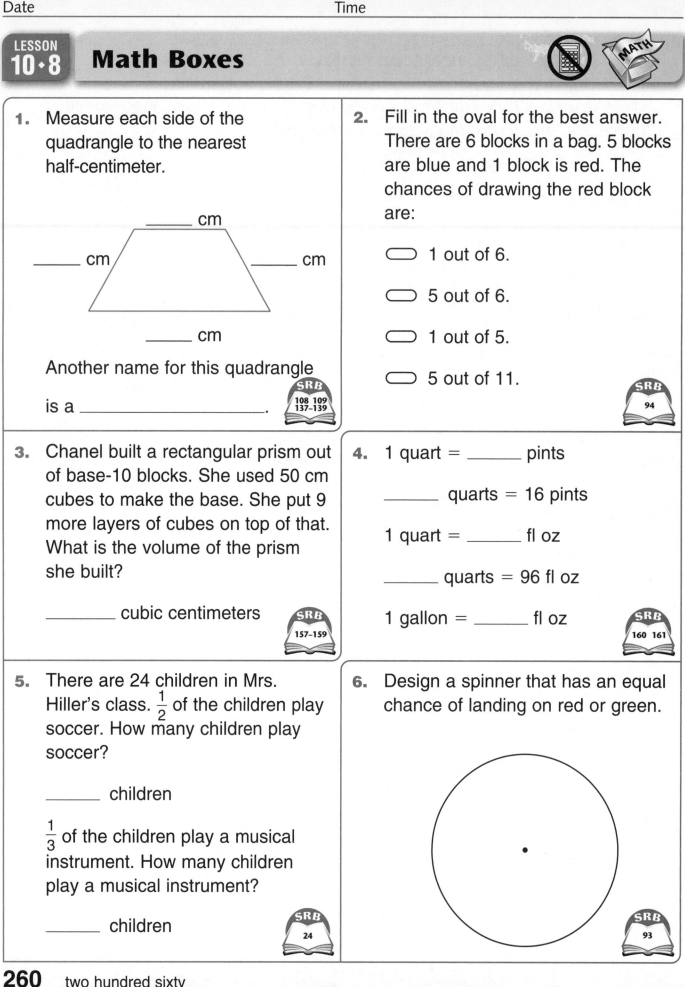

SRB 93

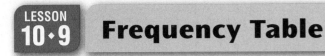 **LESSON 10·9** **Frequency Table**

1. Fill in the table of waist-to-floor measurements for the class.
   This kind of table is called a frequency table.

Waist-to-Floor Measurement (inches)	Frequency	
	Tallies	Number
Total =		

2. What is the median (middle value) of the measurements? _____ in.

3. What is the mean (average) of the measurements? _____ in.

4. The *mode* is the measurement, or measurements, that occur most often. What is the mode of the waist-to-floor measurements for the class? _____ in.

LESSON 10·9 **Bar Graph**

Make a bar graph of the data in the frequency table on journal page 261.

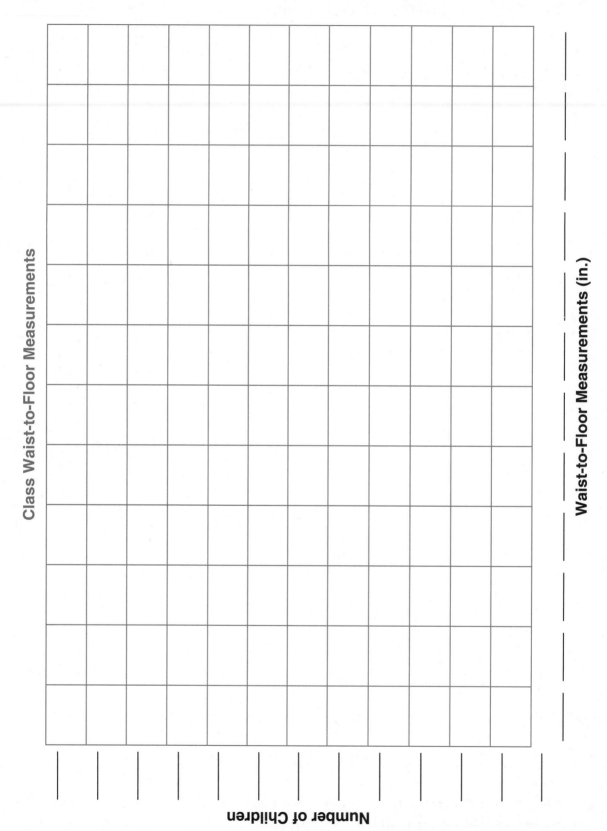

Class Waist-to-Floor Measurements

Waist-to-Floor Measurements (in.)

Number of Children

LESSON 10·9   Math Boxes

1. Use the partial-products algorithm to solve.

   $\begin{array}{r} 83 \\ \times\ 44 \\ \hline \end{array}$ $\begin{array}{r} 72 \\ \times\ 36 \\ \hline \end{array}$

   SRB 68 69

2. 1 pint = _____ fluid ounces

   _____ pints = 48 fluid ounces

   1 half-gallon = _____ quarts

   _____ half-gallons = 6 quarts

   1 liter = _____ milliliters

   SRB 160 161

3. Jerry has 16 fish in a tank. Color $\frac{3}{8}$ of the fish red, $\frac{1}{4}$ of the fish blue, and the rest yellow. What fraction of the fish are yellow?

   _____

   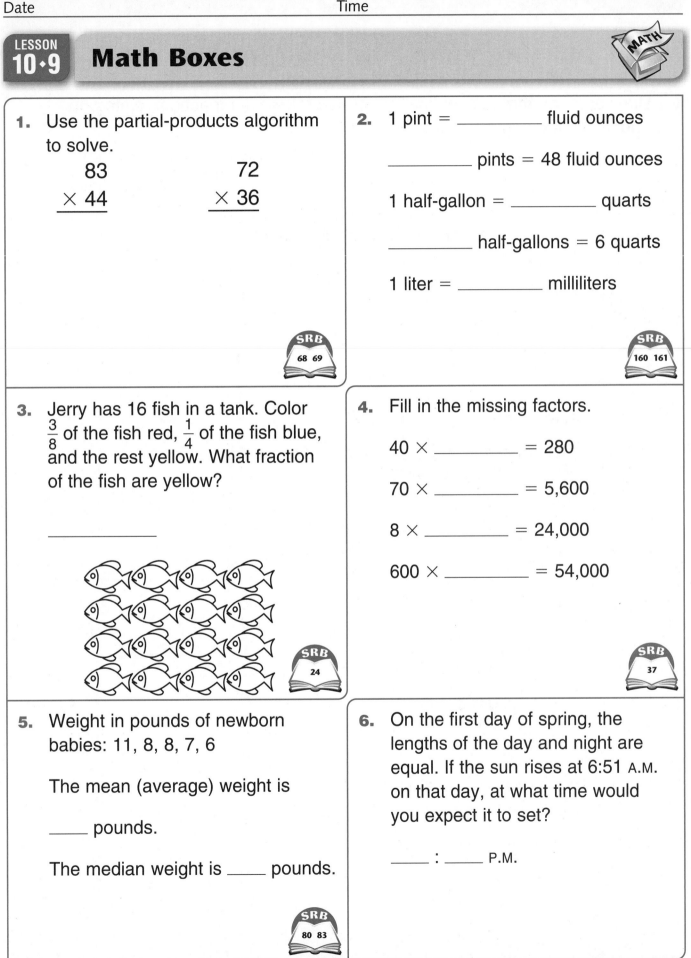

   SRB 24

4. Fill in the missing factors.

   $40 \times$ _____ $= 280$

   $70 \times$ _____ $= 5{,}600$

   $8 \times$ _____ $= 24{,}000$

   $600 \times$ _____ $= 54{,}000$

   SRB 37

5. Weight in pounds of newborn babies: 11, 8, 8, 7, 6

   The mean (average) weight is

   _____ pounds.

   The median weight is _____ pounds.

   SRB 80 83

6. On the first day of spring, the lengths of the day and night are equal. If the sun rises at 6:51 A.M. on that day, at what time would you expect it to set?

   _____ : _____ P.M.

LESSON 10·10 Plotting Points on a Coordinate Grid

1. Draw a dot on the number line for each number your teacher dictates.
Then write the number under the dot.

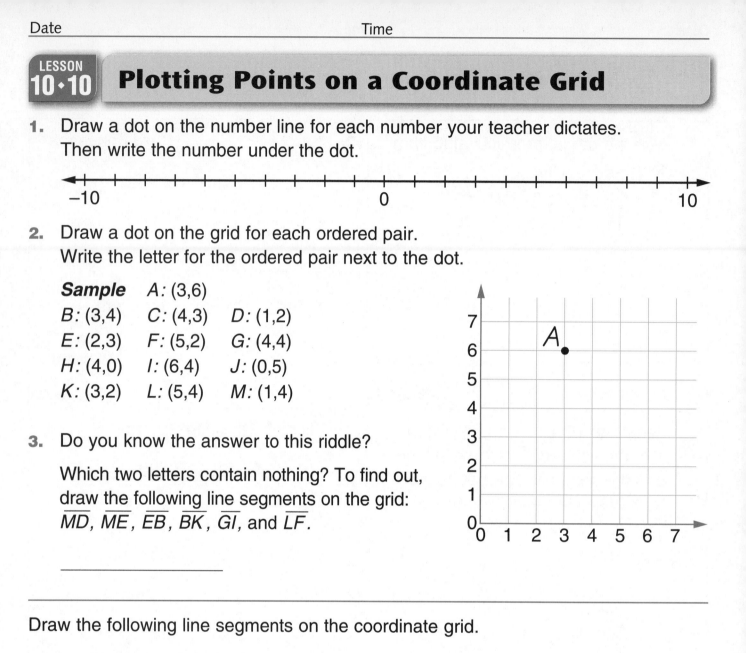

```
←─┼──┼──┼──┼──┼──┼──┼──┼──┼──┼──┼──┼──┼──┼──┼──┼──┼──┼──┼──┼──→
  −10                        0                              10
```

2. Draw a dot on the grid for each ordered pair.
Write the letter for the ordered pair next to the dot.

Sample   *A:* (3,6)
B: (3,4)   *C:* (4,3)   *D:* (1,2)
E: (2,3)   *F:* (5,2)   *G:* (4,4)
H: (4,0)   *I:* (6,4)   *J:* (0,5)
K: (3,2)   *L:* (5,4)   *M:* (1,4)

3. Do you know the answer to this riddle?

Which two letters contain nothing? To find out,
draw the following line segments on the grid:
$\overline{MD}$, $\overline{ME}$, $\overline{EB}$, $\overline{BK}$, $\overline{GI}$, and $\overline{LF}$.

_____

Draw the following line segments on the coordinate grid.

4. From (0,6) to (2,7); from (2,7) to (3,5);
from (3,5) to (1,4); from (1,4) to (0,6)

What kind of quadrangle is this?

_____

5. From (7,0) to (7,4); from (7,4) to (5,3);
from (5,3) to (5,1); from (5,1) to (7,0)

What kind of quadrangle is this?

_____

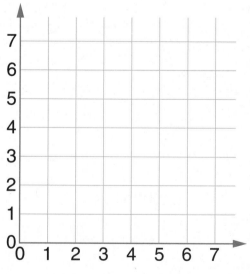

LESSON 10·10 Math Boxes

1. Use the partial-products algorithm to solve.

$$82 \times 35 \qquad 94 \times 76$$

SRB 68 69

2. There are 4 quarts in a gallon. How many quarts of paint did Sally use if she used $1\frac{1}{2}$ gallons of paint? Fill in the circle next to the best answer.

○ **A** 4 quarts

○ **B** 6 quarts

○ **C** 8 quarts

○ **D** 16 quarts

SRB 160 161

3. There are _____ books in $\frac{2}{5}$ of a set of 25 books.

There are _____ minutes in $\frac{3}{4}$ of an hour.

I have six books. This is $\frac{1}{6}$ of a set of books. How many books are in the complete set?

_____ books

SRB 24

4. 50 is 10 times as much as

_____.

700 is _____ times as much as 7.

_____ is 100 times as much as 90.

60,000 is 1,000 times as much as

_____.

5. Number of fish caught each weekend at Aunt Mary's lake:

3, 6, 5, 1, 7, 1, 5

The median number of fish

caught: _____

The mean (average) number

of fish caught: _____

SRB 80 83

6. Anchorage, Alaska has very long days in the summer. In the middle of July, the sun rises at about 3:20 A.M. and sets at about 10:20 P.M. About how many hours of daylight are there?

About _____ hours

LESSON 10·11 Math Boxes

1. What is the mode of the test scores for the class? _____ %

Test Score	Number of Children
100%	///
95%	/////
90%	///// ///
85%	////

SRB 81

2. It is 7:45 A.M. Draw the hour and minute hands to show the time 15 minutes earlier. What time does the clock show now? _____

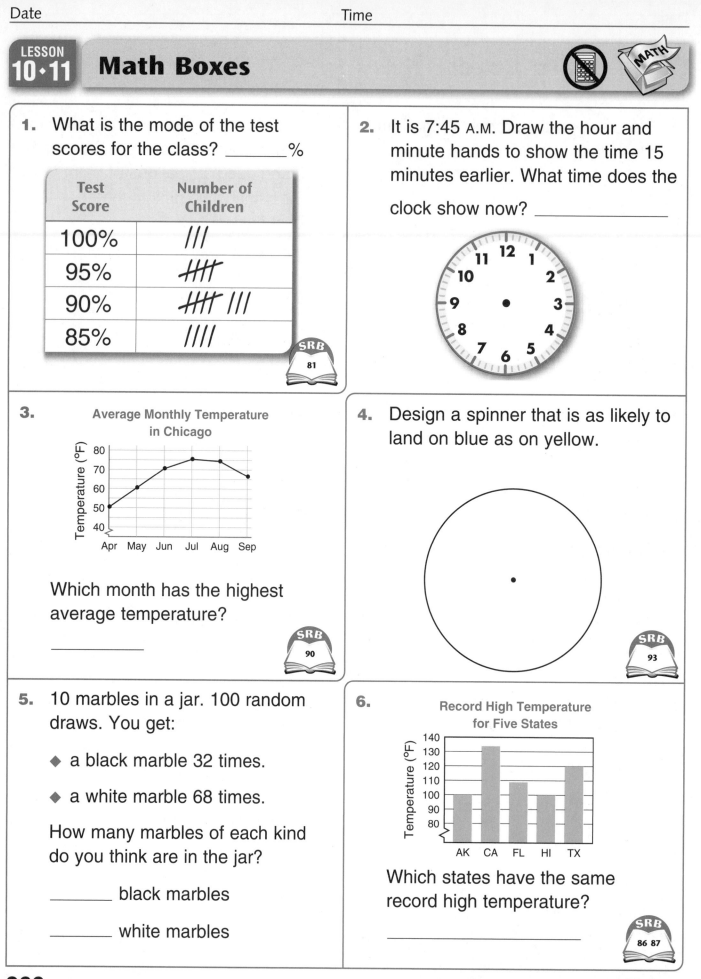

3. **Average Monthly Temperature in Chicago**

Which month has the highest average temperature?

_____

SRB 90

4. Design a spinner that is as likely to land on blue as on yellow.

SRB 93

5. 10 marbles in a jar. 100 random draws. You get:

◆ a black marble 32 times.

◆ a white marble 68 times.

How many marbles of each kind do you think are in the jar?

_____ black marbles

_____ white marbles

6. **Record High Temperature for Five States**

Which states have the same record high temperature?

_____

SRB 86 87

LESSON 11·1 Math Boxes

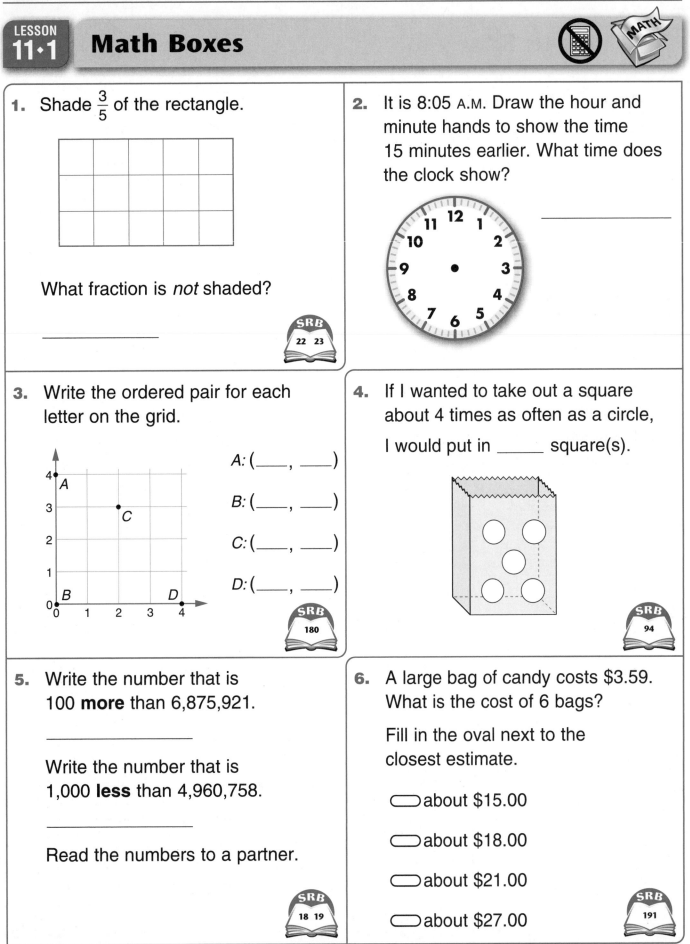

1. Shade $\frac{3}{5}$ of the rectangle.

What fraction is *not* shaded?

_____

SRB
22 23

2. It is 8:05 A.M. Draw the hour and minute hands to show the time 15 minutes earlier. What time does the clock show?

_____

3. Write the ordered pair for each letter on the grid.

A: (____, ____)

B: (____, ____)

C: (____, ____)

D: (____, ____)

SRB
180

4. If I wanted to take out a square about 4 times as often as a circle, I would put in _____ square(s).

SRB
94

5. Write the number that is 100 **more** than 6,875,921.

_____

Write the number that is 1,000 **less** than 4,960,758.

_____

Read the numbers to a partner.

SRB
18 19

6. A large bag of candy costs $3.59. What is the cost of 6 bags?

Fill in the oval next to the closest estimate.

⬭ about $15.00

⬭ about $18.00

⬭ about $21.00

⬭ about $27.00

SRB
191

LESSON 11·2 Math Boxes

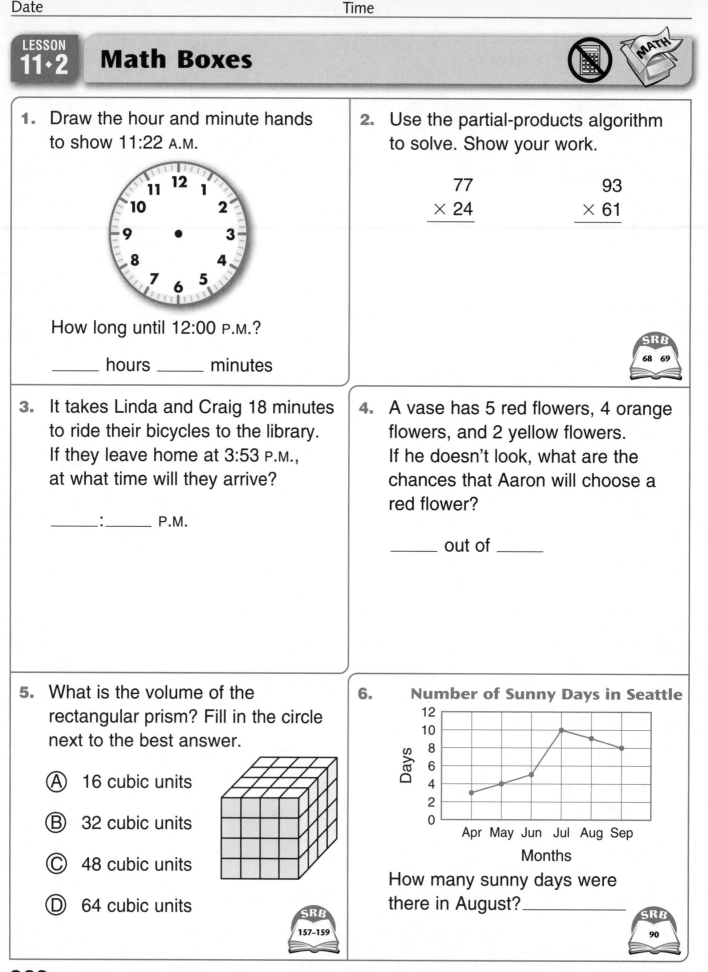

1. Draw the hour and minute hands to show 11:22 A.M.

How long until 12:00 P.M.?

_____ hours _____ minutes

2. Use the partial-products algorithm to solve. Show your work.

$$\begin{array}{r} 77 \\ \times\ 24 \\ \hline \end{array} \qquad \begin{array}{r} 93 \\ \times\ 61 \\ \hline \end{array}$$

SRB
68 69

3. It takes Linda and Craig 18 minutes to ride their bicycles to the library. If they leave home at 3:53 P.M., at what time will they arrive?

_____ : _____ P.M.

4. A vase has 5 red flowers, 4 orange flowers, and 2 yellow flowers. If he doesn't look, what are the chances that Aaron will choose a red flower?

_____ out of _____

5. What is the volume of the rectangular prism? Fill in the circle next to the best answer.

Ⓐ 16 cubic units

Ⓑ 32 cubic units

Ⓒ 48 cubic units

Ⓓ 64 cubic units

SRB
157–159

6. **Number of Sunny Days in Seattle**

Days (y-axis: 0, 2, 4, 6, 8, 10, 12)
Months (x-axis: Apr, May, Jun, Jul, Aug, Sep)

How many sunny days were there in August?_____

SRB
90

268 two hundred sixty-eight

Date _____ Time _____

LESSON 11·3 Spinners

Math Message

Color each circle so that it matches the description.

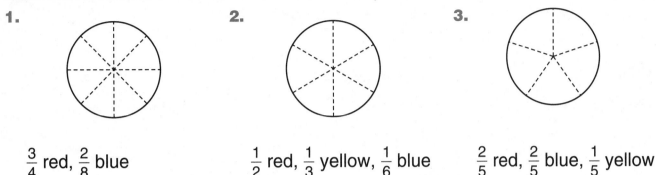

1.

$\frac{3}{4}$ red, $\frac{2}{8}$ blue

2.

$\frac{1}{2}$ red, $\frac{1}{3}$ yellow, $\frac{1}{6}$ blue

3.

$\frac{2}{5}$ red, $\frac{2}{5}$ blue, $\frac{1}{5}$ yellow

Spinner Experiments

Tape *Math Masters,* p. 367, to your desk or table.
Make a spinner on the first circle.

4. Spin the paper clip 10 times. Tally the
number of times the paper clip lands on
the shaded part and on the white part.

Lands On	Tallies
shaded part	
white part	

5. Record results for
the whole class.

Lands On	Totals
shaded part	
white part	

Make a spinner on the second circle.

6. Spin the paper clip 10 times. Tally the
number of times the paper clip lands on
the shaded part and on the white part.

Lands On	Tallies
shaded part	
white part	

7. Record results for
the whole class.

Lands On	Totals
shaded part	
white part	

8. With the second spinner, the paper clip has a better chance of landing on the

_____ part of the spinner than on the _____ part.

LESSON 11·3 Estimate, Then Calculate

For each problem, make a ballpark estimate and circle the phrase that best describes your estimate. Next, calculate the exact sum or difference. Check that your answer is close to your estimate.

1. more than 500

less than 500

$$\begin{array}{r} 825 \\ -\ 347 \\ \hline \end{array}$$

2. more than 500

less than 500

$$\begin{array}{r} 984 \\ -\ 392 \\ \hline \end{array}$$

3. more than 500

less than 500

$$\begin{array}{r} 658 \\ -\ 179 \\ \hline \end{array}$$

4. more than 500

less than 500

$$\begin{array}{r} 227 \\ +\ 285 \\ \hline \end{array}$$

5. more than 500

less than 500

$$\begin{array}{r} 324 \\ +\ 161 \\ \hline \end{array}$$

6. more than 500

less than 500

$$\begin{array}{r} 179 \\ +\ 338 \\ \hline \end{array}$$

LESSON 11·3 · **Math Boxes**

1. How many thirds are shaded?

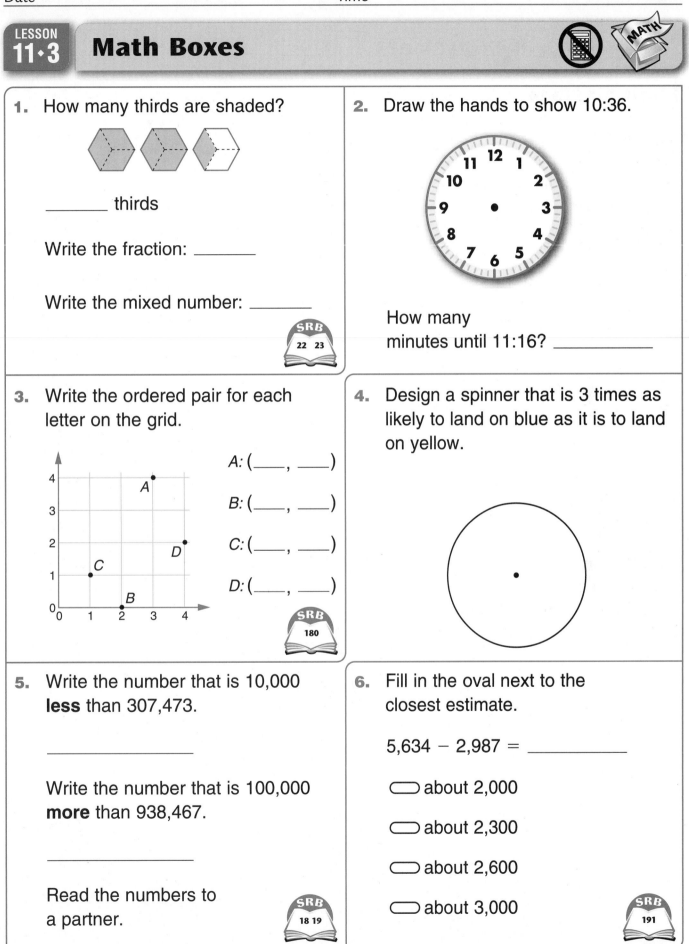

_____ thirds

Write the fraction: _____

Write the mixed number: _____

SRB
22 23

2. Draw the hands to show 10:36.

How many
minutes until 11:16? _____

3. Write the ordered pair for each letter on the grid.

A: (____, ____)

B: (____, ____)

C: (____, ____)

D: (____, ____)

SRB
180

4. Design a spinner that is 3 times as likely to land on blue as it is to land on yellow.

5. Write the number that is 10,000 **less** than 307,473.

_____

Write the number that is 100,000 **more** than 938,467.

_____

Read the numbers to
a partner.

SRB
18 19

6. Fill in the oval next to the closest estimate.

$5,634 - 2,987 =$ _____

⬭ about 2,000

⬭ about 2,300

⬭ about 2,600

⬭ about 3,000

SRB
191

LESSON 11·4 Making Spinners

Math Message

1. Use exactly six different colors. Make a spinner so the paper clip has the **same chance** of landing on any one of the six colors.

   (*Hint:* Into how many equal parts should the circle be divided?)

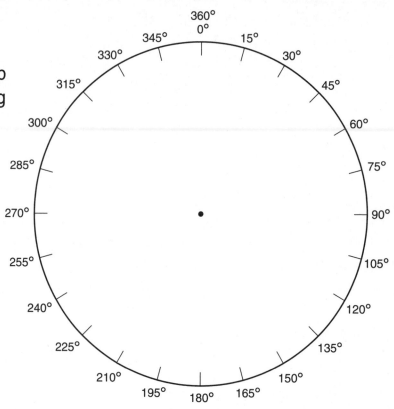

2. Use only blue and red. Make a spinner so the paper clip is **twice as likely** to land on blue as it is to land on red.

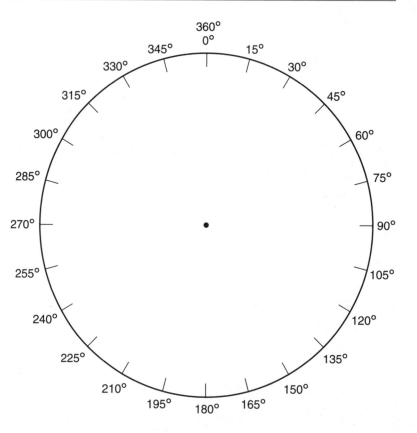

LESSON 11·4

Making Spinners *continued*

3. Use only blue, red, and green. Make a spinner so the paper clip:

 ◆ has the **same chance** of landing on blue and on red

and

 ◆ is **less likely** to land on green than on blue.

4. Use only blue, red, and yellow. Make a spinner so that the paper clip:

 ◆ is **more likely** to land on blue than on red

and

 ◆ is **less likely** to land on yellow than on blue.

LESSON 11·4 Math Boxes

1. Draw the hands to show 9:34 A.M.

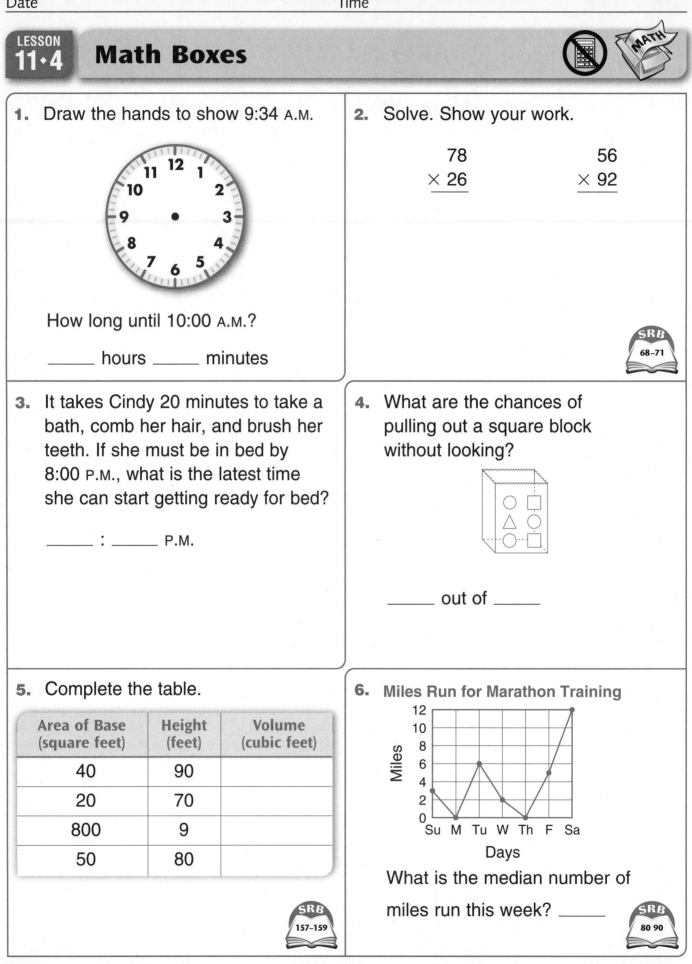

How long until 10:00 A.M.?

_____ hours _____ minutes

2. Solve. Show your work.

$$\begin{array}{r} 78 \\ \times\ 26 \\ \hline \end{array} \qquad \begin{array}{r} 56 \\ \times\ 92 \\ \hline \end{array}$$

SRB
68–71

3. It takes Cindy 20 minutes to take a bath, comb her hair, and brush her teeth. If she must be in bed by 8:00 P.M., what is the latest time she can start getting ready for bed?

_____ : _____ P.M.

4. What are the chances of pulling out a square block without looking?

_____ out of _____

5. Complete the table.

Area of Base (square feet)	Height (feet)	Volume (cubic feet)
40	90	
20	70	
800	9	
50	80	

SRB
157–159

6. Miles Run for Marathon Training

What is the median number of miles run this week? _____

SRB
80 90

LESSON 11·5 Random-Draw Problems

Each problem involves marbles in a jar. The marbles are blue, white, or striped. A marble is drawn at random (without looking) from the jar. The type of marble is tallied. Then the marble is returned to the jar.

◆ Read the description of the random draws in each problem.

◆ Circle the picture of the jar that best matches the description.

1. From 100 random draws, you get:

a blue marble ● 62 times.

a white marble ○ 38 times.

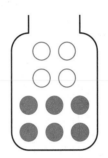

10 marbles in a jar

10 marbles in a jar

2. From 100 random draws, you get:

a blue marble ● 23 times.

a white marble ○ 53 times.

a striped marble ⊘ 24 times.

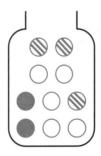

10 marbles in a jar

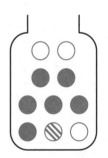

10 marbles in a jar

Try This

3. From 50 random draws, you get:

a blue marble ● 30 times.

a white marble ○ 16 times.

a striped marble ⊘ 4 times.

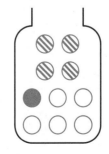

10 marbles in a jar

10 marbles in a jar

LESSON 11·5 Reading and Writing Numbers

Write the value of 7 for each column below.

L		K	J		I	H	G		F	E	D	.	C	B	A
hundred millions	,	ten millions	millions	,	hundred thousands	ten thousands	thousands	,	hundreds	tens	ones	.	tenths	hundredths	thousandths
7	,	7	7	,	7	7	7	,	7	7	7	.	7	7	7

Example: Column K: _70,000,000 or 70 millions_

1. Column A: _____

2. Column G: _____

3. Column F: _____

4. Column I: _____

5. Column C: _____

6. Column B: _____

7. Column L: _____

Write the numbers that your teacher dictates.

8. _____

9. _____

10. _____

11. _____

12. _____

13. _____

LESSON 11·5   **Math Boxes**

---

1. Jessica makes $3.25 every hour she works at the lemonade stand. Will she make enough money to buy a $12 watch if she works from 10:30 A.M. to 2:30 P.M.?

_____

---

2. Number of books read during summer: 9, 9, 5, 15, 3, 9, 6

The mean number of books read is _____.

The median number of books read is _____.

SRB
80
83–85

---

3. Write the number that has

3 in the hundred-thousands place,

6 in the thousands place,

4 in the ten-thousands place,

1 in the millions place, and

5 in all of the other places.

____ , ____ ____ ____ , ____ ____ ____

SRB
18 19

---

4. Color the spinner so it is more likely to land on blue than orange, and more likely to land on green than blue.

---

5.   **2-Hour Water Polo Practice**

30 min scrimmage

25 min swimming warm up

10 min passing

25 min goal shots

30 min team drills

Which two activities together make up $\frac{1}{2}$ of a 2-hour water polo practice?

_____ and _____

---

6. Danielle skates from 6:45 to 7:30 every morning and from 3:05 to 3:55 in the afternoon on Mondays and Wednesdays. How long does she skate in a week?

____ hours ____ minutes

---

two hundred seventy-seven   **277**

LESSON 11·6   **Math Boxes**

1. Fill in the oval next to the closest estimate.

    747 + 932 = _____

    ⬭ about 1,500

    ⬭ about 1,700

    ⬭ about 2,000

    ⬭ about 2,500

    SRB
    191

2. What is the median number of pets children have? _____ pet(s)

Number of Pets	Number of Children
0	///
1	~~HHT~~ ~~HHT~~
2	~~HHT~~
3	///
4	/
5	/

    What is the mode number of pets? _____ pet(s)

    SRB
    76
    80 81

3. Write the number that has

    1 in the ten-thousands place,

    7 in the thousands place,

    2 in the hundred-thousands place,

    8 in the millions place, and

    0 in all of the other places.

    ____, ____ ____ ____, ____ ____

    SRB
    18 19

4. Design a spinner that is twice as likely to land on blue as it is to land on yellow.

5. **Weights of 5 Dogs**

    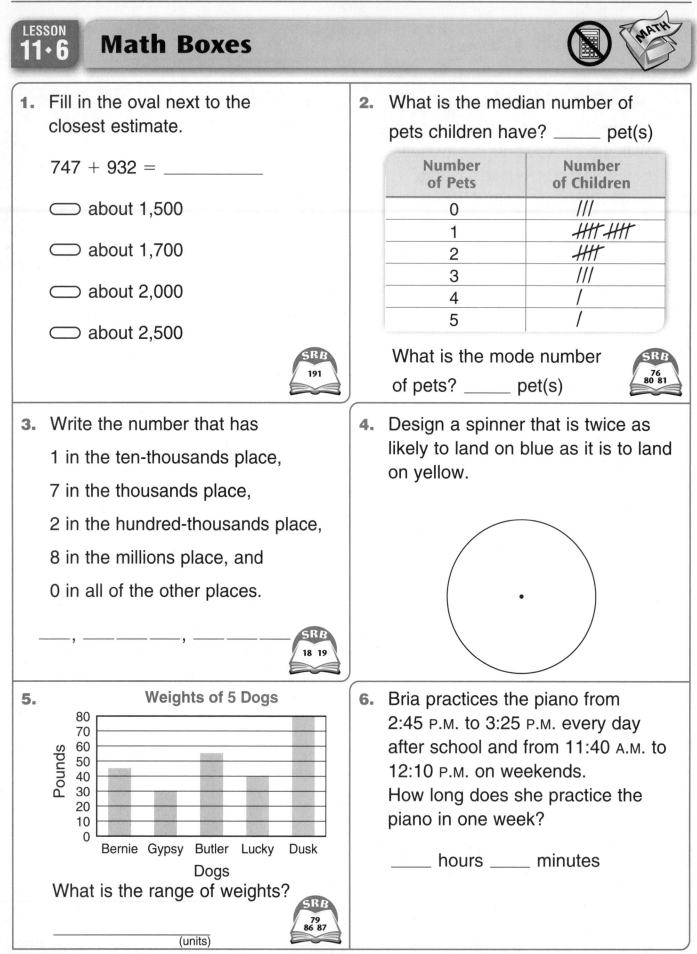

    Bernie Gypsy Butler Lucky Dusk
    Dogs

    What is the range of weights?

    _____
    (units)

    SRB
    79
    86 87

6. Bria practices the piano from 2:45 P.M. to 3:25 P.M. every day after school and from 11:40 A.M. to 12:10 P.M. on weekends. How long does she practice the piano in one week?

    _____ hours _____ minutes

LESSON 11·1

Sunrise and Sunset Record

Date	Time of Sunrise	Time of Sunset	Length of Day
			hr min
			hr min
			hr min
			hr min
			hr min
			hr min
			hr min
			hr min
			hr min
			hr min
			hr min
			hr min
			hr min
			hr min
			hr min
			hr min
			hr min
			hr min
			hr min
			hr min
			hr min

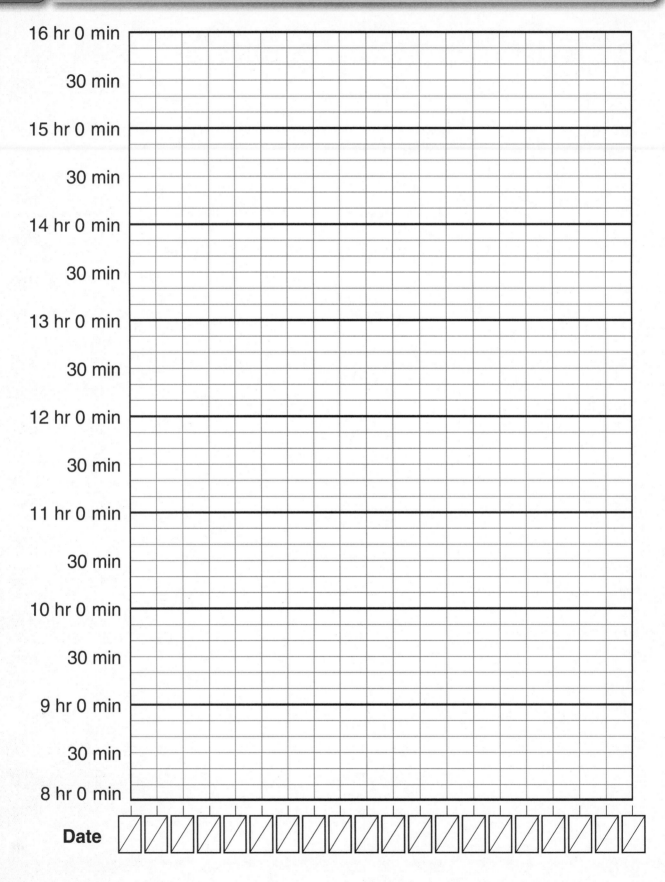

LESSON 11·1

Length-of-Day Graph

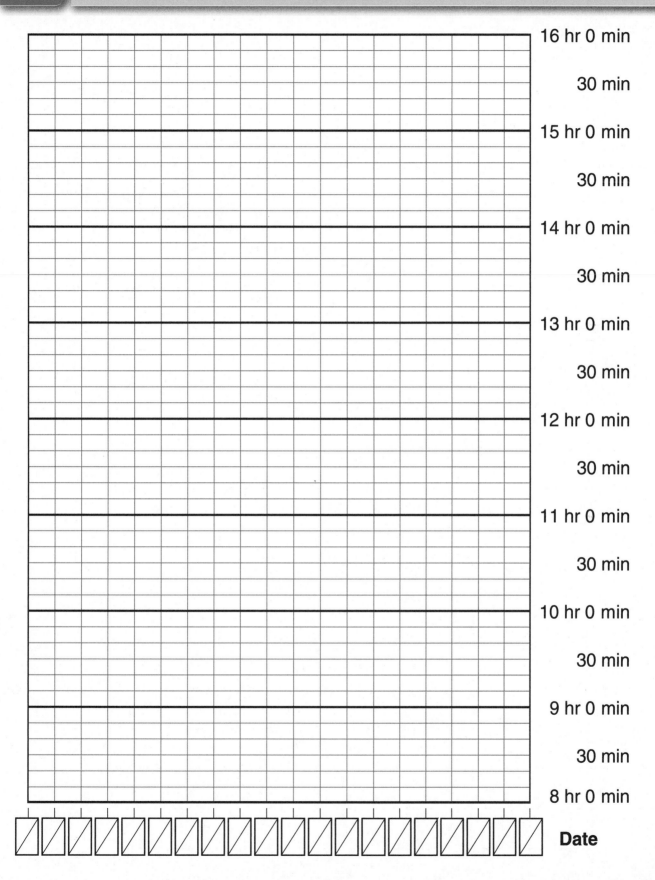

16 hr 0 min

30 min

15 hr 0 min

30 min

14 hr 0 min

30 min

13 hr 0 min

30 min

12 hr 0 min

30 min

11 hr 0 min

30 min

10 hr 0 min

30 min

9 hr 0 min

30 min

8 hr 0 min

Date

Date   Time

Notes

Date _____ Time _____

Notes

_____

_____

_____

_____

_____

_____

_____

_____

_____

_____

_____

_____

_____

_____

_____

_____

_____

_____

_____

LESSON 8·5 Fraction Cards

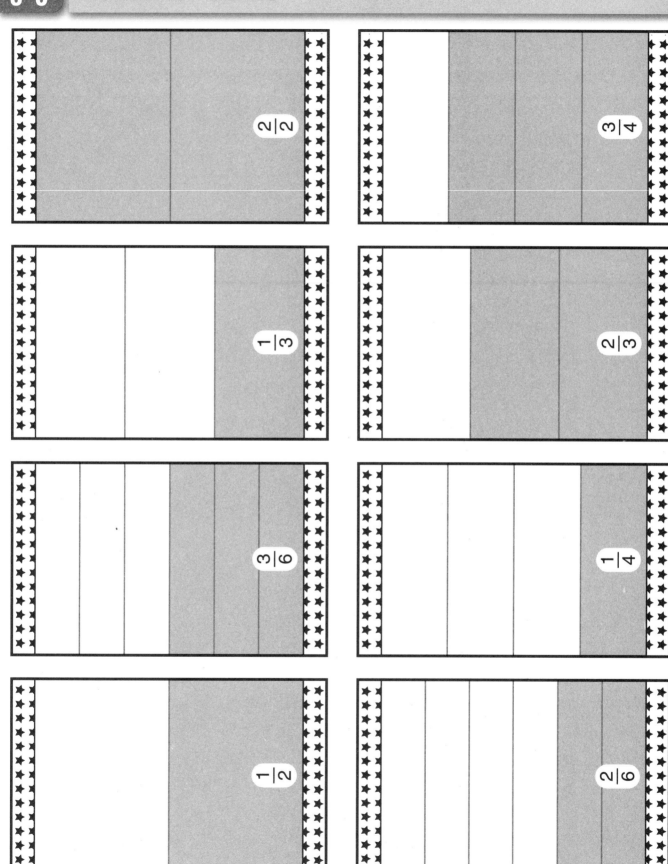

Activity Sheet 5

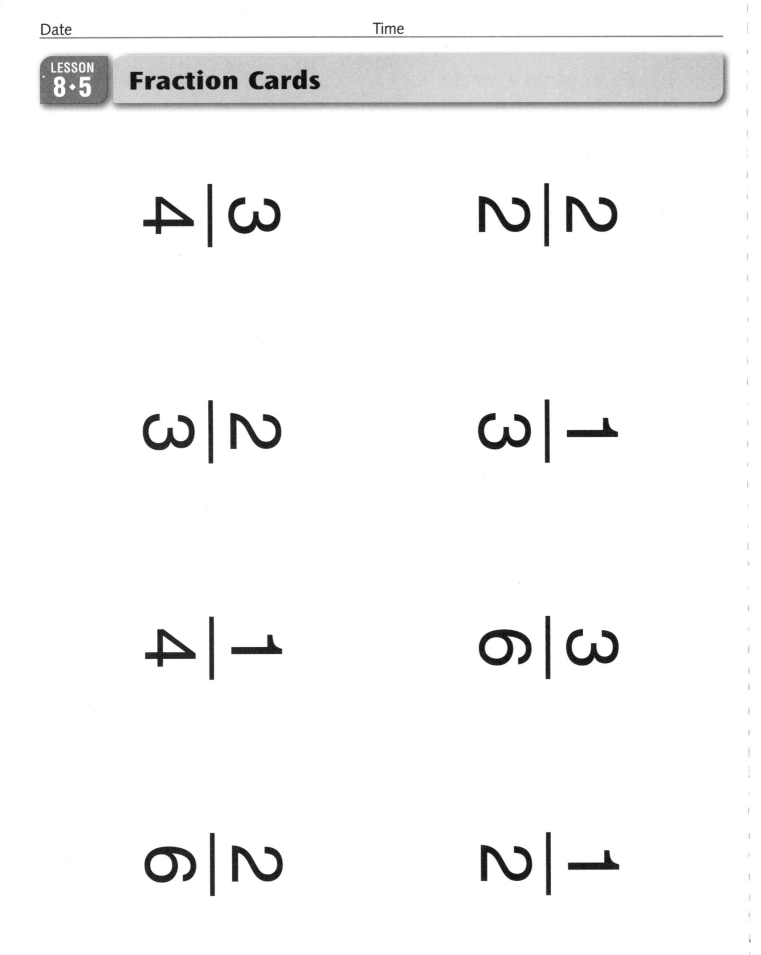

Back of Activity Sheet 5

LESSON 8·5

Fraction Cards

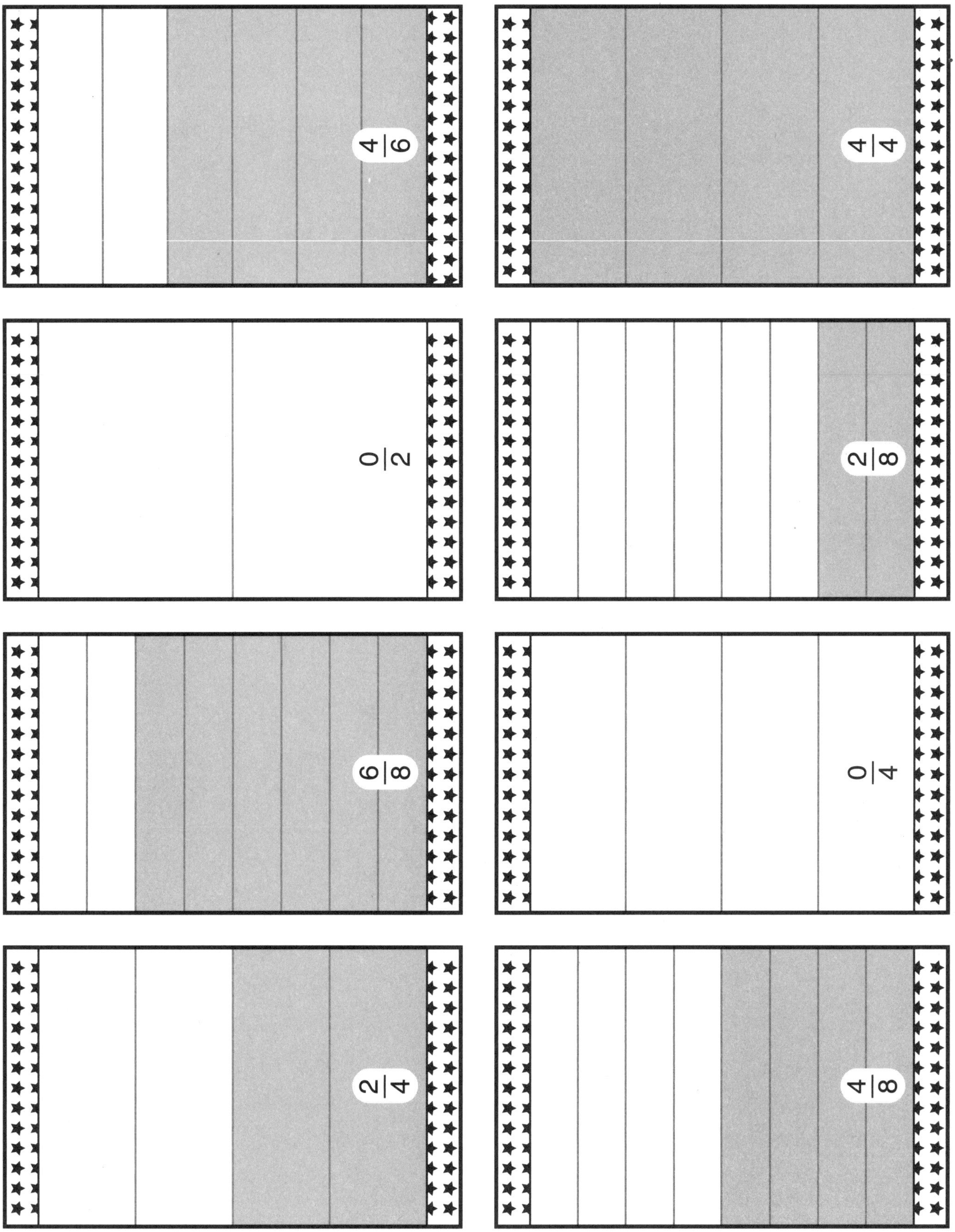

$\dfrac{4}{6}$

$\dfrac{4}{4}$

$\dfrac{0}{2}$

$\dfrac{2}{8}$

$\dfrac{6}{8}$

$\dfrac{0}{4}$

$\dfrac{2}{4}$

$\dfrac{4}{8}$

Activity Sheet 6

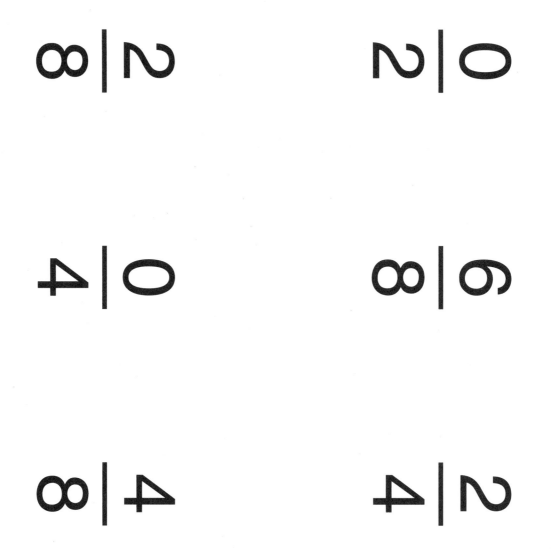

Date _____ Time _____

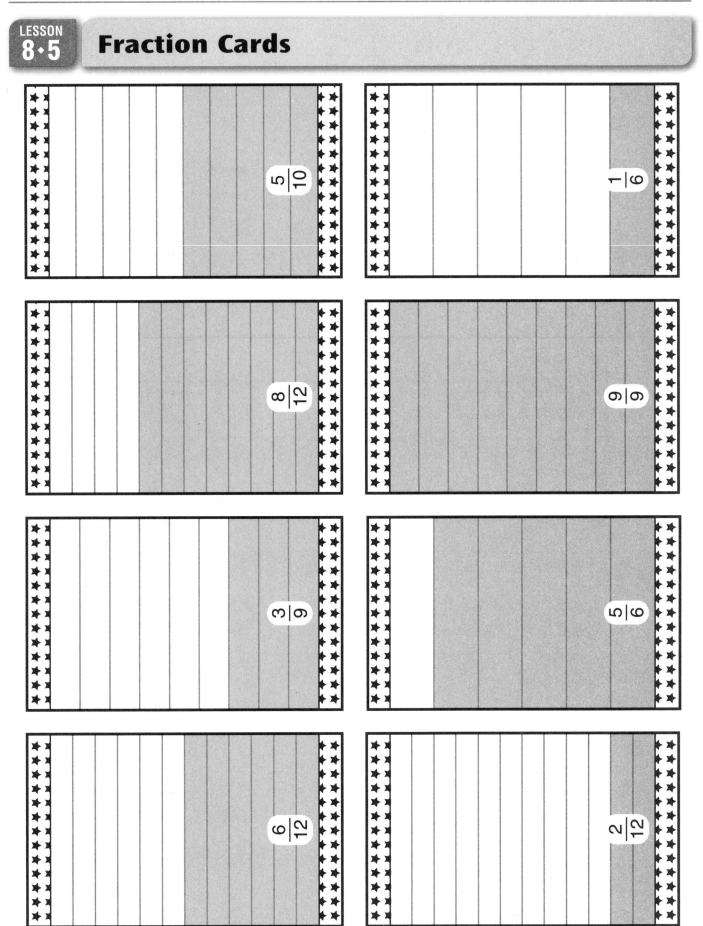

$\frac{5}{10}$

$\frac{1}{6}$

$\frac{8}{12}$

$\frac{9}{9}$

$\frac{3}{9}$

$\frac{5}{6}$

$\frac{6}{12}$

$\frac{2}{12}$

Activity Sheet 7

LESSON 8·5 Fraction Cards

$$\frac{5}{10} \qquad \frac{1}{6}$$

$$\frac{8}{12} \qquad \frac{9}{9}$$

$$\frac{3}{9} \qquad \frac{5}{6}$$

$$\frac{6}{12} \qquad \frac{2}{12}$$

Back of Activity Sheet 7

LESSON 8·5 Fraction Cards

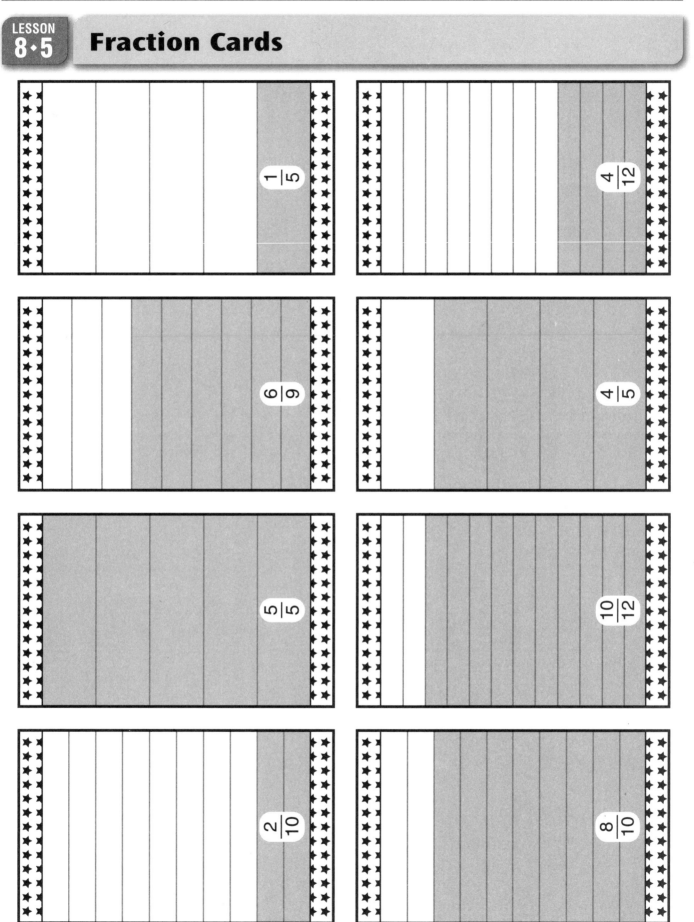

LESSON 8·5 Fraction Cards

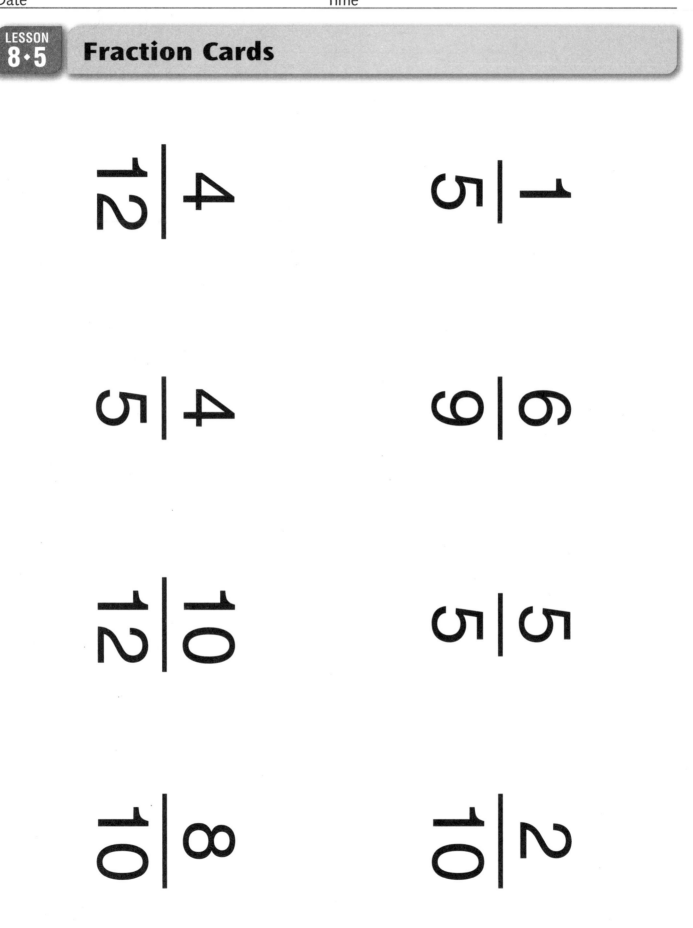

$$\frac{1}{5} \qquad \frac{4}{12}$$

$$\frac{6}{9} \qquad \frac{4}{5}$$

$$\frac{5}{5} \qquad \frac{10}{12}$$

$$\frac{2}{10} \qquad \frac{8}{10}$$

Back of Activity Sheet 8